Lewin Jürgen Rohde

AF874178

The Universal Sea Language

Being a Complete Code of Signals for Day and Night

Translated from the Danish by Captain H.B. Dahlerup

Elibron Classics
www.elibron.com

Elibron Classics series.

© 2007 Adamant Media Corporation.

ISBN 0-543-69577-8 (paperback)
ISBN 0-543-69576-X (hardcover)

This Elibron Classics Replica Edition is an unabridged facsimile of the edition published in 1835 by A. W. Webster, London.

Elibron and Elibron Classics are trademarks of Adamant Media Corporation. All rights reserved.

This book is an accurate reproduction of the original. Any marks, names, colophons, imprints, logos or other symbols or identifiers that appear on or in this book, except for those of Adamant Media Corporation and BookSurge, LLC, are used only for historical reference and accuracy and are not meant to designate origin or imply any sponsorship by or license from any third party.

THE

Universal Sea Language;

BEING

A COMPLETE

CODE OF SIGNALS

for Day and Night.

ADAPTED TO THE USE

OF

VESSELS OF ALL NATIONS,

AND REQUIRING

NO ADDITIONAL FLAGS,

OR MEANS, BUT SUCH AS ARE FOUND IN EVERY VESSEL, EVEN THE SMALLEST FISHING CRAFT.

BY

LEVIN JOERGEN ROHDE,

CAPTAIN IN THE ROYAL DANISH NAVY; CAPTAIN OF THE PORT OF ST. THOMAS, WEST INDIES; KNIGHT OF THE DANNEBROG.

PUBLISHED ALSO IN DANISH, FRENCH, AND GERMAN.

TRANSLATED FROM THE DANISH

BY CAPTAIN H. B. DAHLERUP, R.D.N.

SECOND EDITION.

LONDON:

PUBLISHED BY A. W. WEBSTER, 156, REGENT STREET;

AND SOLD BY

R. B. BATE, CHART AGENT TO THE ADMIRALTY, 21, POULTRY,

E. WILSON, BOOKSELLER, 88, ROYAL EXCHANGE,

AND AT EVERY PORT IN THE UNITED KINGDOM.

1835.

C. WHITING, BEAUFORT HOUSE, STRAND.

AGENTS

OF WHOM THE CODE CAN BE PROCURED AT THE OUT-PORTS.

ABERDEEN	Messrs. A. Brown and Co.
ABERYSTWITH	Mr. Josiah Evans.
BELFAST	Mr. George Phillips.
BERWICK	Mr. Andrew Reid.
BRIGHTON	Mr. R. Loder.
BRISTOL	Mr. John Braham.
CHATHAM	Mr. James Burrill.
CHESTER	Messrs. Poole and Boult.
CORK	Mr. J. Bolster.
——	Mr. H. Bennett.
DEAL	Mr. Thos. Hayward.
DEVONPORT	Mr. W. C. Cox.
DUBLIN	Messrs. Milliken and Son.
——	Mr. F. C. Spear.
DUNDEE	Mr. Fred. R. Shaw.
FALMOUTH	Mr. Alfred B. Duckham.
——	Mr. James Lake.
GLASGOW	Messrs. John Smith and Son.
GLOUCESTER	Mr. J. E. Lea.
GRAVESEND	Messrs. Talbot and Co.
GREENOCK	Messrs. Heron and M'Gregor.
HULL	Mr. Isaac Wilson.
LANCASTER	Mr. Hudson.
LEITH	Messrs. W. Reid and Son.
LIVERPOOL	Messrs. Bywater, Dawson, & Co.
LYNN	Mr. Bullen.
MILFORD HAVEN	Mr. E. Pritchard.
MONMOUTH	Mr. Charles Hough
NEWCASTLE	Messrs. Currie and Bowman.
NORWICH	Mr. J. Fletcher.
POOLE	Mr. John Sydenham.
PORTSMOUTH	Mr. George Stebbing.
RAMSGATE	Mr. M. Labatt.
SHEERNESS	Mr. W. Knewstub.
STOCKTON	Mr. Thos. Jennett.
SOUTHAMPTON	Messrs. J. R. and H. Stebbing.
SUNDERLAND	Messrs. Reed and Son.
WEYMOUTH	Mr. John Commins.
WHITEHAVEN	Mr. J. W. Callander.
YARMOUTH (GREAT) . . .	Mr. Wm. Alexander.

ADVERTISEMENT.

The following description of the peculiar advantages of this Code of Signals having been obligingly presented to the Publisher by Captain Sir John Ross, by whom it was communicated to the British Association for the Advancement of Science, he feels he cannot introduce the work to the nautical world with a recommendation better calculated to obtain for it a favourable reception, Sir John having been signal-lieutenant to the Baltic Fleet, under the command of Admiral Lord de Saumarez, for several years.

Copy of a Paper presented at the British Association for the Advancement of Science, assembled at Dublin on the 14th of August, 1835.

"UNIVERSAL SEA LANGUAGE;"

being a complete system of communications between the crews of ships of different nations, without any knowledge of each other's language.

This ingenious and simple Code of Signals was first communicated to me by the gallant inventor, Captain Rohde, of the Royal Danish Navy, at Copenhagen, in July, 1834, and in September last I had the honour of submitting the English MS. to our excellent King, who having perused it with attention, commanded me to transmit it to the Lords Commissioners of the Admiralty; at the same time expressing his high approbation of the system. Here, however, it met with some delay, probably from the changes which took place in that Board, and from the circumstance that no less than three codes of signals were at the same time under the consideration of their lordships, and it was not until April last, when I had an interview with Lord Auckland, that their report on this interesting subject was obtained: this could not but be favourable, and the usual number of copies were subscribed for; as also by the Honourable East India Company, the Committee of Lloyd's, the Corporation of the Trinity House, &c. &c.; and at length the English edition is in progress; the French edition, which is also a translation from the original Danish, has been already printed, the Government having subscribed for 200 copies. The German, Spanish, and Dutch translations are soon to follow. The advantages of this method of communication by signals over every other, are briefly these:

In the first place, it will be found by far the cheapest, the whole expense being the price of the book, which is only sixteen shillings; the purchase of flags or other symbols being unnecessary.

Secondly: The only materials required on board any ship are the flags under which she sails, Jack, Ensign, and Pendant (the colour being immaterial), and two white flags, for which two table cloths, or if there are none on board, two shirts, or any thing that will make a flag will suffice; so that whatever is required is already to be found even in the smallest craft. By these simple and ready means, communications of any and of every kind may be made by an English vessel to a foreign one, and *vice versâ*, without the least knowledge of each other's language, and under circumstances of peril and distress which have rendered every other mode impracticable.

Again: Those on the sea-coast, who would wish to save their fellow-creatures from a watery grave, might point out to the stranger an unknown harbour or creek, or the best place to run on shore, and by these invaluable signals, convey to a perishing crew of any nation, every information required to assist their humane endeavours; while on the other hand, the crew of a stranded ship might convey to the spectators of their perilous situation every thing that is requisite, even the perishing foreigner's last farewell to relatives and friends.

I can safely assure the Section that during my services in His Majesty's Navy, of above forty years, had I been in possession of these signals, and had they been generally distributed and published, as they are now intended to be, I should have witnessed the saving of hundreds of lives, and thousands of pounds in valuable property.

JOHN ROSS,
Captain in the Royal Navy.

It is only necessary to add the following documents from Public Bodies in approbation of the work, or ordering Copies.

DOCUMENTS REFERRED TO.

TRANSLATED FROM THE DANISH.

No. I.

LETTER FROM THE BOARD OF ADMIRALTY.

Copenhagen, 15th May, 1834.

SIR,

IT is with great pleasure that the undersigned Board declare, by these presents, that they regard the System of Signals for the use of Vessels of all Nations, of which you are the Author, and which is herewith returned, as being extremely useful, and entirely answering the purpose for which you have designed it.

College of the Admiralty and the Commissariat,

(Signed) H. HOLSTEN. KINCK.
ROTHE. WIBORG. STIERNHOLM, *Secretary.*

To Captain L. J. Rohde, Captain of the Port of St. Thomas,
Knight of the Order of Dannebrog.

No. II.

LETTER FROM THE ROYAL BOARD OF CUSTOMS.

Copenhagen, 31st May, 1834.

SIR,

On returning the manuscript of your System of Signals for Vessels of all Nations, it is with pleasure the Board of Customs informs you, that on investigation of its general utility and importance, and particularly in reference to the care and skill with which the Code is completed, it can do no otherwise than desire that it should be generally adopted.

The Royal Board of Customs and Commerce,

(Signed) LOUZOW. C. MALLING.
LEHMANN. WEDEL.
THONNING. GARLIEB.
H. BECH. SCHLEGEL, *Secretary.*

To Captain of the Royal Navy,
L. J. Rohde.

No. III.

LETTER FROM THE ROYAL PRIVILEGED MARINE INSURANCE COMPANY.

Copenhagen, 19th June, 1834.

SIR,

Together with your letter, dated the 4th inst., you have been pleased to send to the undersigned Directors "A Code of Signals for Vessels of all Nations," invented by you. In hereby returning you this your work, the undersigned Directors feel a pleasure in informing you, that in reference to the want, hitherto existing, of a complete and universally-intelligible medium of communication at sea, it is their opinion that the publishing and general use of your work will be of great benefit to commerce and navigation in general, and of peculiar advantage to Marine Insurance in particular.

The Directors of the Royal Privileged Marine Insurance Company,

	Treschow.	Joerg. Bech, jun.
(Signed)	P. Moller.	B. Wigelsen.
	G. Hagerup.	J. D. Meyer.

To Captain of the Royal Navy, L. J. Rohde,
Knight of Dannebrog.

A true translation of the original certificates laid before me.

Witness my hand and official seal. Fr. Gandil,
Translator and Interpreter, by Royal authority duly admitted and sworn.

Copenhagen, 3d July, 1834.

TRANSLATED FROM THE FRENCH.

No. IV.

Letter from the Minister of Marine and the Colonies.

Paris, 28th March, 1835.

Sir,

The Committee to whom I referred for examination the Code of Signals, of which you are the Author, has rendered me a very favourable account of the work; and representing that its adoption would be generally useful to navigators, has expressed its wish that it should be patronised by the Government of France.

In consequence of this recommendation I have determined that there shall be purchased on account of my Department, 200 copies of the work.

I also feel myself bound to inform the Minister for the Department of Commerce, of the importance attached to the publication, that he may also order a certain number of Copies for his Department, as soon as its general adoption by other countries shall be established.

I beg, Sir, you will thus perceive the interest I take in your work, and my desire to extend its usefulness to the whole body of nautical persons, for whose use it is designed.

I have the honour herewith to return you the manuscript, that you may immediately proceed to its publication.

Receive the assurance of my particular consideration.

(Signed) Admiral, Peer of France, Minister of the Marine and the Colonies, Duperré.

Captain Rohde,
His Danish Majesty's Service, at Paris.

No. V.

LETTER FROM THE MINISTER OF COMMERCE.

Paris, 15th May, 1835.

SIR,

In consequence of the favourable testimonials with which you have furnished me from the Committee appointed by the Minister of Marine, who have examined your System of Signals, I have decided to order for my Department fifteen copies of the work you are about to publish.

I intend these copies for distribution to the different Chambers of Commerce at our ports, who will not fail to appreciate their usefulness, and feel bound to impress their importance on the officers of the mercantile marine.

You will have the goodness to send these copies to the Office of Manufactures of my Department, Rue de Grenelle, Saint Germaine, No. 122; and as soon as the books are received, I will give the necessary orders for their payment.

Receive, Sir, the assurance of my high esteem.

(Signed) The Minister Secretary of State for Commerce,

DUCHATEL.

Capt. L. J. Rohde,
Rue Pinon, No. 22, à Paris.

I, the underwritten interpreter and translator, duly sworn and admitted, do hereby certify, that the preceding copies and the translations are exact, and conform to the originals exhibited unto me, and which I have signed *ne varietur*.

Witness my hand and seal of office,

D. MELDOLA.

Paris, Rue Richelieu, No. 115,
25th May, 1835.

ENGLISH.

No. VI.

LETTER FROM THE COURT OF DIRECTORS OF THE HONOURABLE EAST INDIA COMPANY.

East India House, 20th June, 1835.

SIR,

Having laid before the Court of Directors of the East India Company your letter, dated the 5th instant, together with the Code of Signals invented by Captain Rohde, which accompanies it, I am commanded to inform you, that the Court have resolved to subscribe for forty copies of the English edition, which it is intended to publish of the Code in question.

I am, Sir, Your most obedient humble servant,

P. AUBER, *Secretary*.

Mr. A. W. Webster,
Publisher for Captain Rohde.

No. VII.

LETTER FROM THE LORDS COMMISSIONERS OF THE ADMIRALTY.

Admiralty, 7th July, 1835.

SIR,

Having laid before my Lords Commissioners of the Admiralty your letter of the 19th of last month, requesting that some copies of a new Code of Signals by Captain Rohde, of the Danish navy, may be taken for the use of His Majesty's ships; I am commanded by their Lordships to acquaint you, that my Lords will take thirty copies of this work, to issue to some ships in His Majesty's navy, in order that their captains may report on its probable utility.

I am, Sir,

Your most obedient humble servant,

C. Wood, *Secretary*.

To Mr. Webster,
Publisher, 156, Regent Street.

No. VIII.

Letter from the Committee of Lloyd's.

Lloyd's, 15th July, 1835.

Sir,

Having laid before the Committee for managing the Affairs of Lloyd's your letters of the 7th and 11th instant, I am directed to inform you, that they will take ten copies of Captain Rohde's Code of Signals, with a view of distributing them to certain experienced masters of vessels to report thereon.

I am, Sir,

Your obedient servant,

Wm. Dobson, *Secretary*.

To Mr. A. W. Webster,
156, Regent Street.

No. IX.

Letter from the Corporation of the Trinity House.

Trinity House, London, 28th August, 1835.

Sir,

I have laid before the Elder Brethren of this Corporation your letter of the 17th instant, with the French translation of Captain Rohde's Code of Marine Signals, and the small portion of that work which has been completed in English; and I am desired to communicate to you the wish of the Board to be furnished with six copies of the English translation.

I am, Sir, your very obedient servant,

J. Herbert, *Secretary*.

To Mr. A. W. Webster.

No. X.

Letter from the Hudson's Bay Company.

Hudson's Bay House, 22d October, 1835.

Mr. Smith, (*Secretary*), requests Mr. Webster will set down the Company as subscribers for two copies of Captain Rohde's Book of Signals.

AUTHOR'S PREFACE.

It will hardly be thought necessary that we should enter largely upon the great utility of a Code of Signals being adopted in common by all civilized nations, as a medium of holding conversation upon the sea. Suffice it here to say, that a Code of Signals arranged for sentences, thus adopted, and understood by the seafaring people of all, or most, of the civilized nations, would in fact substitute a universal language, enabling every captain of a vessel to converse with another met with at sea, of whatever nation, though both are strangers to each other's tongue.

It is a well-known fact, that few Spaniards, Italians, or Frenchmen, are familiar with the northern languages; and if now and then one is met with, who speaks the English language,—which among seafaring people of the northern nations is pretty generally understood and spoken,—it commonly happens, that his pronunciation is so incorrect, as to make a conversation through the speaking-trumpet quite unintelligible, and misinterpretations and mistakes of the most serious consequence are likely to take place.

Among the obstacles to the introduction of a universal language at sea, by means of a Code of Signals adopted by the merchantmen of all nations, may unquestionably be reckoned the great number of flags and pendants hitherto considered requisite; which is not only a considerable expense, but the system is at the same time seemingly rendered too complex to persons unversed in this branch of naval service. To obviate these objections, the present Code of Signals has been arranged upon such principles, as to fit it for the use of every merchant-vessel, without burthening the owners with even the least extra expense in the outfit, as the flags and pendant required are no more than what are commonly found in even the smallest craft; when at the same time it is hoped, that a needless complexity has been avoided, though a vast number of signals are obtained from the simplest means. The signal flags required are, namely:

1. National flag or colours.
2. A second flag, or jack of any colour, white excepted (the quarantine flag for example).
3. A broad pendant, or a narrow long piece of whatever stuff or colour, only differing in shape from a common flag.

4 & 5. Two white flags, or, in want of them, a pair of white sheets or table cloths, or even a pair of white shirts may be substituted.

It is by the transposition of these five signals, upon principles, it is believed, never before used, that a very considerable and sufficient number of signals are procured.

As another essential reason, why a universal Code of Signals has not yet been generally adopted, this may be advanced: That the authors of the different proposed, and (with the exception of the abovementioned imperfections) well planned, Codes of Signals, have omitted to have them translated into different languages at once. It is only when the great public (the seafaring people of all nations) has a new invention put into its hands fully prepared for its immediate use, that it can be expected to become generally adopted, and it is only in this way that its utility can be properly valued by all. On the strength of this reason, the author of this Code of Signals has taken care to have it published simultaneously in English Danish, French, and German, each signal signifying the same sentence in the aforementioned four languages. That such a translation into other languages is practicable only with sentence signals, while it is utterly impossible with a telegraphic system, which is formed of words and syllables, is a thing well known to all practised persons.

The author being moved by no interested views, save an ardent desire to be useful in his profession, has spared no pains whatever, in order, as much as lay in his power, to pave the way for a general adoption of these signals. He has had his plan laid before the governments and respective authorities of the principal seafaring nations, submitting to their wisdom to take such measures, as, suiting peculiar forms and circumstances, might be thought most advisable in order to promote a free and cheerful adoption of what may, perhaps within a very short time, prove itself of great utility to the seafaring world.

It would be vanity to pretend that the plan, hereby laid before the public, is practically faultless; or to lay claim to much merit in the planning of a scheme so simple, and the execution of matters generally so well understood; but the author was struck by the idea as worthy our present times, that difference of language ought no longer to exist between men of a profession that connects the most distant parts of our globe; men, who ploughing the same trackless waste, share the same hardships, the same dangers, and therefore ought to be found ever ready and ever willing, like brethren, to assist each other in word and deed. If, therefore, the author's opinion, backed as it is by that of men, on whose judgment he ought to have reliance, deceive him not, both as to the practicability of the plan itself, and the sufficiency of the proposed means, he thinks himself justified in the hope, that all Shipowners, Societies of Merchants, and Insurance Companies, will lend a helping hand to promote a scheme that tends so much to strengthen the ties of humanity and philanthropy. It is only by their efforts, their efficacious recommendations, each in his sphere, that this Code of Signals, if really it merits their protection, may be expected to become adopted by many at once, and then only its utility will be obvious to all.

Before closing these lines, the Author avails himself of the opportunity of publicly acknowledging the great obligation he is under to his friend Captain H. B. Dahlerup, of the Royal Danish Navy, to whose friendly criticisms, and able assistance in the methodical arrangement of this work, he feels highly indebted.

To Admiral Baron Duperré, Minister of the Marine and the Colonies of France, he begs to renew the expression of his profound and respectful gratitude, for the flattering reception with which he honoured him, and the encouragement he has been pleased to extend to his work.

To Captain Sir John Ross, C.B., of the British Royal Navy, the intrepid Arctic navigator, is also due his thanks for the favourable notice he procured for his book, both of his Sovereign, and the Board of Admiralty, and whose testimony, from his long acquaintance with the subject of signals, renders his approbation the more flattering.

GENERAL CONTENTS.

II.

THE

GREAT CODE OF SENTENCES,

EXPRESSED BY

PAIR SIGNALS.

PREPARATORY SIGNAL.

THIS signal is made with the view of ascertaining whether the vessel addressed is in possession of this present Code of Signals, and consequently it invites the other to enter into conversation; or more properly, it signifies that the Great Code of Signals *by pairs* is to be used in the ensuing conversation. The vessel thus addressed acknowledges to be in possession of these signals, and consequently to have understood the signal made, by the hoisting of a *white flag*.

The white flag is always made use of for the answering of a signal, or signifying the signal made to be observed and understood; and though, generally, *two* successive signals are wanted to express a sentence, *each* of them are to be answered *separately*, taking care always to haul down the answering flag when the signal goes down.

When it happens that a signal cannot be properly distinguished, either on account of its being hoisted from a bad place, or the flags not blowing out clear, the following signal is hoisted instead of the answering flag, signifying

"Signal flying not to be distinguished."

Note.—To make the flags show clear in calm weather, a light piece of wood may be tied to the upper edge or leech of each flag, as a kind of yard; and a topping lift made fast near the outer end of the lowermost, and tied, at proper distances, to the other flag-yards; so that, when the flags are hoisted, they may be brought in a horizontal position by topping in the lift.

INTRODUCTION

TO

THE GREAT CODE.

The present System, or Code of Signals, is constructed upon the principle of *transposition of Signals by pairs*. In this manner, to express a sentence, *two successive signals* are required, the numbers of which, as belonging to them in the following signal table, are noted down in the columns before the respective sentences. The signal table itself contains forty signals, formed of the *different transposition of the flags*.

A number is given to each sentence, according to the place or numerical order it occupies; this number is found in the first of the columns, and does not interfere with the signal numbers.

The signal flags, as before mentioned, are:

1. A broad pendant, or a long and tapering piece of bunting, or other light stuff, of any colour.

2 & 3. Two coloured flags, national or others.

4 & 5. Two white flags, or in want of these, two table cloths, sheets, or such like, provided the colour be but white.

Proper Names

are signalized in single letters, each letter having its proper single signal assigned to it in the signal table, from No. 1, signifying A, No. 2, B, and so forth, to the end of the alphabet of each language. But to indicate that a succession of signals are to be understood as single ones, and denoting letters of the alphabet, the pendant alone is first hoisted; and the same ought to be observed at the beginning of every new name or word, should more follow, without interruption of other signals. To denote the finishing of a name or word, where there is only one, or when it is the last of several others signalized in this manner, the signal for the closing letter when answered, is first hauled half-mast, or nearly down, then hoisted aloft a second time, and finally struck.

The Sign of Abbreviation.

As it will be found desirable, in signalizing a word or proper name in the above-mentioned manner, to abbreviate as much as may be done, without the risk of being misunderstood, a single signal, No. 29 (') denoting the sign of abbreviation, has been added to the single signals for letters of the alphabet.

The following example will show how words or proper names may be abbreviated.

The name of a vessel is to be signalized. Let it be "The Duke of Wellington."

Then hoist: 1. Pendant alone, indicating single signals, denoting letters to follow.
2. Single signal No. 4, denoting letter D.
3. ditto No. 29, denoting sign of abbreviation.
4. ditto No. 23, signifying letter W.

and so forth, through all the letters composing the name WELLINGTON.

Note.—The article *the*, and the preposition *of*, are omitted as superfluous.

All personal names are given in the above manner: the Christian name, when required, is signalized first, either the first letter alone, or where mistakes may be apprehended, more, or all, of its component letters; and then the family or surname, which it will generally be necessary to give without abbreviation.

Saint, or *St.*

As the name of a person or town might happen to begin with this word, a single signal, No. 30, has likewise been given to it along with the letters of the alphabet.

For example: ST. ANTHONY is signalized by hoisting,

1. Pendant alone.
2. No. 30, signifying SAINT.
3. No. 1, signifying the letter A, and so on.

It is, perhaps, superfluous to remark, that the names of towns and places, not to be found in the Compound Table (of which hereafter), must be spelt in the above manner.

Numbers and Fractions

are signalized as follows:

Whole numbers, according to the directions given in page 81, for the use of the Compound Table; but if there be fractions, they ought always to follow *after* the whole numbers, and in order to give notice that such are to come, the white flag is hoisted alone first, as a distinguishing signal; next follows a single signal, whose number in the signal table is that of the *Numerator*, and then follows the single signal whose number denotes the *Denominator*.

For example: the fraction $\frac{3}{4}$ is to be signalized:

Hoist 1. *The White Flag* indicating fractions to come.
2. The single signal, No. 3.
3. ditto No. 4.

In this manner may be signalized all the fractions which may be formed out of the numbers from 1 to 40; consequently, fractions formed of greater numbers must be reduced, as nearly as possible, to a denominator not exceeding 40.

A Man is Overboard

is signalized by hoisting and hauling down the national flag or colour as quickly as possible, and continuing the same, until it is observed by the other vessels lying near, which of course are expected to render immediate assistance. This signal has been preferred to others, on the supposition, that in a case like this, where assistance may be expected, and can be of any avail (in harbour, or in company with other vessels), the colours are likely to be already up or bent; and it is therefore the easiest and quickest mode of indicating an accident of this kind, by hauling the colours up and down until observed.

It ought to be borne in mind that in the present code, the common measure for distances is *the league*, of which 20 go to a degree; it is of course equal to 3 miles.

When a vessel is making signals, where a number of other vessels are lying together, it ought to be a rule, that the nearest one is meant, which ought to consider it a duty to answer the questions made, and give the information desired. It is only on the presumption of a liberality of spirit like this, which shuns not a trouble, trifling in itself, that the idea, which gave birth to the present code, is built; every man ought at least to consider, that if he to-day refuses to render a trifling service to a brother in profession, he has no right to claim the same for himself to-morrow or another day.

As it may, however, not unfrequently happen, that it is desired to exchange signals with a certain vessel out of several lying together (a countryman or an acquaintance, for example), the following rule, to be observed in an instance like this, may in some degree serve to make up for the want of particular signals or private numbers, the introduction of which, into a *universal* code of signals, will be allowed to be impracticable.

When it is desired to make signals to a certain vessel lying among several others, this vessel is pointed out in the following manner: The bearing of the vessel is taken, and a signal hoisted to indicate the point of compass in which she lies from the vessel making signals; however, in this particular case, the common mode of signalizing points of the compass is not made use of; but the single signals in the signal table are used to express the number of the point of compass in question, reckoning from North through East and South, &c.; North being No. 1; N by E, No. 2; NNE, No. 3; and so forth round the compass to N by W, answering to No. 32. At the same time one of the remaining flags, or the only one left at disposal, is kept flying from another masthead or place. As this is the only instance in this code of signals, where two signals are hoisted at the same time from different places in the ship, it cannot fail to attract notice, and every one of the surrounding vessels may easily, by reversing the bearings signalized, make out if she be the vessel meant or not; when, in the former case, the signal is repeated in the usual way, upon which all are hauled down, and the signalizing goes on in the usual manner.

XXII. SIGNAL TABLE FOR

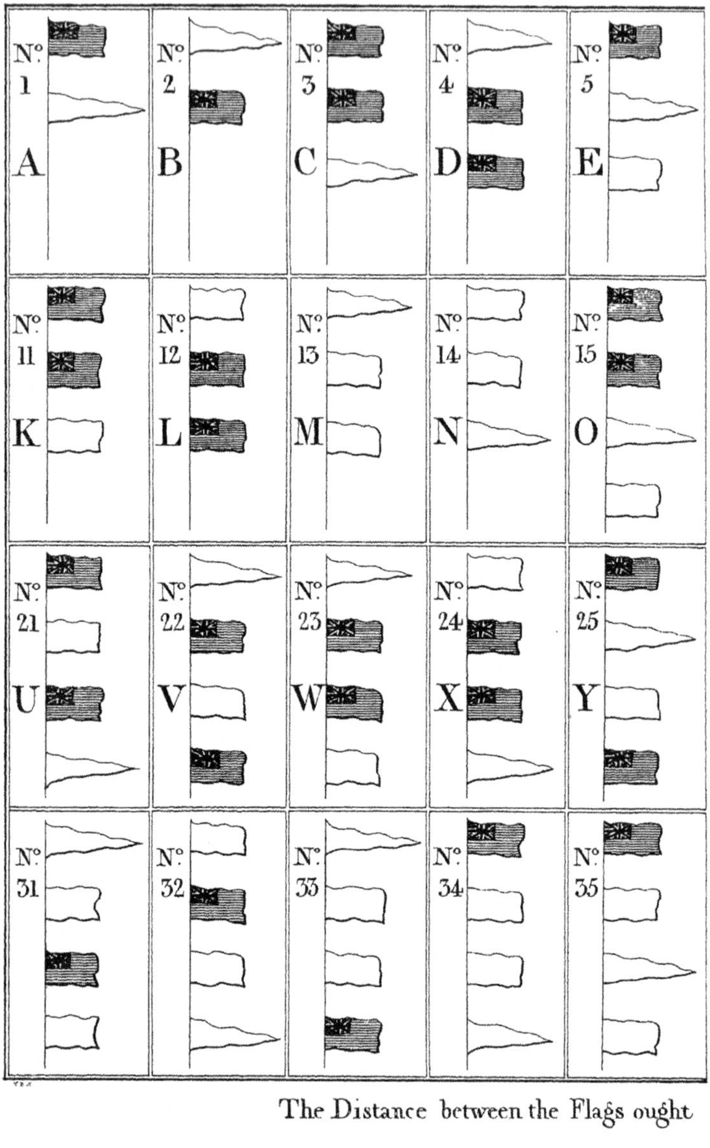

The Distance between the Flags ought

THE GREAT CODE. XXIII.

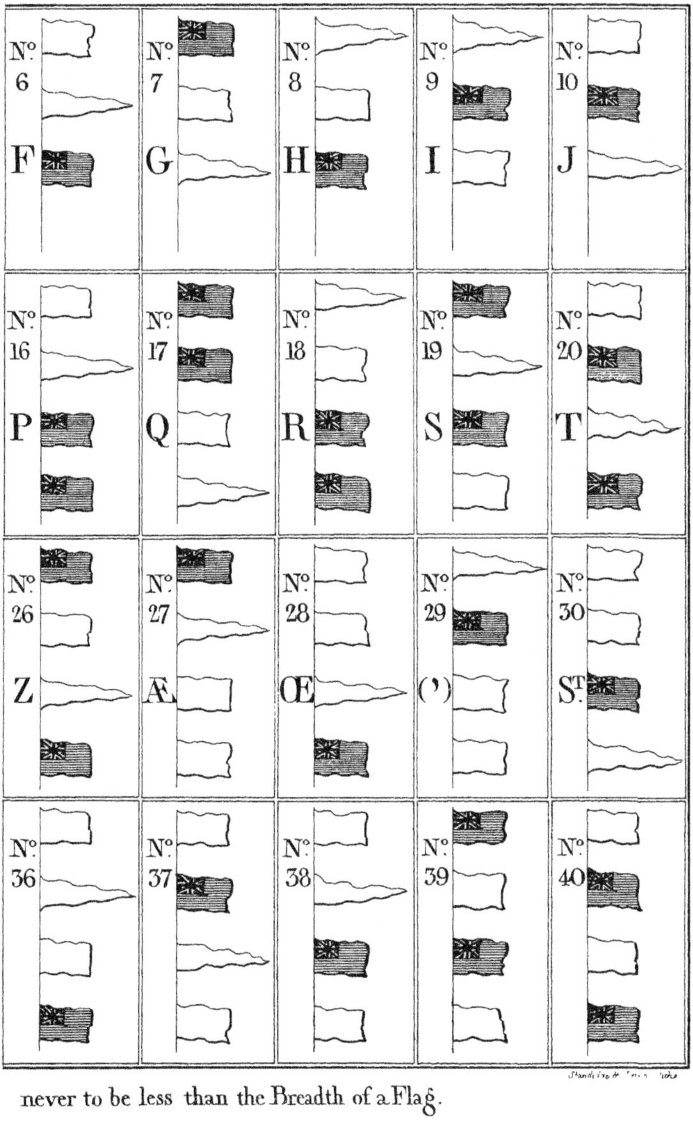

never to be less than the Breadth of a Flag.

EXPLANATION

OF THE

GREAT CODE OF SIGNALS.

THE sentences are arranged in chapters, with a superscription showing the matter treated of, and these again are subdivided into two parts, namely, Questions, and Answers or Communications. An Index for the facilitating the use of the code of sentences is constructed upon this principle, of bringing the matter under different heads, and these subdivided again as above mentioned. It will, however, often be necessary to have recourse for answers to the Compound Table, which will be explained hereafter (see page 79).

The following observations will be found to confer all the illustration the subject will admit of.

The use of the Code of Sentences will be found quite plain and easy, and even the finding out the sentence belonging to a pair of signals made, will occasion no great difficulty, as it is only necessary, after having noted down the numbers of both signals, to search the column before the sentences for the first number, and then by running over the following combinations in pairs, which are put down in numerical order, the pair in question will soon be found out.

After the preparatory signal has been interchanged (see page xvii), which on many occasions may be omitted, *the first signal* must always be taken from among the sentences, whether it be interrogatory or communicating. In like manner the answer is given by a sentence, and generally an appropriate answer may be found under the same chapter with the question made; yet it may occasionally be taken from any of the other chapters, but never from the Compound Table, except when that table is expressly indicated in the question made, or the preceding sentence signal. When in the latter case a signal has been answered from the Compound Table, where perhaps it would have been thought more proper to answer it by a sentence, as being more explicit, this may be done afterwards, and such an answer then assumes the character of a communication.

The 16th Chapter has been composed of sentences, which though of frequent occurrence in conversations at sea, yet cannot properly be brought under any of the foregoing chapters. Neither would the heterogeneous character of these sentences admit of their being placed under any distinct head of matter, on account of which the title of "Sentences of a Miscellaneous Nature," has been prefixed to this chapter, which should be studied with diligence, in order to become familiar with its contents.

Finally: the 17th Chapter is formed of a few sentences set apart for *general answers* to sentences, such as "Yes." "No." "I do not know, or my knowledge is uncertain." "I must take time to consider," &c.

The pair signals used for these sentences are purposely omitted in the Compound Table, so that no mistake can possibly take place, when, for instance, a signal has been made, the answer to which is expected from the Compound Table.

To avoid all manner of mistakes, the following rules ought ever strictly to be observed:

1. To answer every signal made by hoisting the white flag.

2. Never to answer a signal till it is clearly made out, and its number noted down.

3. Not to give the answer requested till the signal has been regularly answered by the white flag, and this is hauled down again.

From this it will be clear, that a signal which cannot properly be distinguished, ought not to be answered; but instead of the white flag, the signal for, "Signal flying not to be distinguished," must be hoisted. (See page xvii.)

CONTENTS OF THE CHAPTERS

OF THE GREAT CODE OF SENTENCES.

SENTENCES

OF THE

GREAT CODE OF SIGNALS.

Note.—In the first 40 numbers of the following pair signals, it is to be observed, that the second signal is the same with the first, the flags only being reversed; so that by hauling down the first signal, and reversing it, the second is formed immediately, without a change of flags.

General Remark.—The numbers attached to the numerical order of sentences in the first column are only referred to occasionally in the sentences following, and have no further use.—Answers to questions relative to numbers must be made from the Compound Table.

CHAPTER I.

Numerical order of Sentences.	PAIR SIGNALS. No.	PAIR SIGNALS. No	*Relating to the Navigation of Ships.* QUESTIONS.
			Note.—For the Signal demanding the Latitude and Longitude to the present moment, see *Selected Signals.*
1	1	2	What are your calculations at noon to-day?
2	3	4	Have you a chronometer?
			Note.—Though it has been found necessary to insert the above sentence here, the question about the longitude by chronometer may be made at once by the sentence following, as the other, if he has no chronometer, may answer, "I have none, or I have had none," which is an answer to be found among the sentences for general answers, forming the 17th chapter of this code of sentences.
3	5	6	What is your longitude by chronometer?
			Ans. From Compound Table.
4	7	8	What is the exact time by your chronometer, that I may compare or regulate mine by it, and name the place to whose meridian it is set?
			Ans. From Compound Table, with two pair signals.
			Note.—Before answering this signal, it is necessary to reduce the time shown by the chronometer to the meridian it is set to. In order

Numerical order of Sentences.	Pair Signals No.	Pair Signals No.	CHAPTER I. *Relating to the Navigation of Ships.* QUESTIONS.
			to give the time with due precision, these reductions (the constant, if there be any, and that for the daily acceleration, or retardation) are applied to a time which is about ten minutes later than that actually shown by the chronometer at that moment; which ten minutes are presumed to elapse in making the two signals. This corrected time, to a full minute, as no seconds can be signalized, is taken in the Compound Table, and the first signal hoisted. When this is answered, it is hauled down, and the second signal hoisted, which is kept flying till the moment the chronometer exactly points to the minute signalized, when it is struck with a sudden jerk, to mark the precise moment. It follows, of course, that the other must be very attentive in observing this jerk, and note down the time by his chronometer; he may then afterwards, by looking over the Compound Table, find out the hours and minutes signalized, and making the necessary reductions, compare the two watches together. When the time has thus been given, then follows the last pair-signal (from Compound Table), denoting the place to whose meridian the chronometer is set.
5	9	10	Of what date is your latest observation for longitude? *Ans.* From the Compound Table.
6	11	12	Of what date is your latest observation for latitude? *Ans.* From the Compound Table.
7	13	14	At what date did you take your departure, and from what place? *Ans.* From the Compound Table, and in the same order as the questions; namely, 1st, the date, and 2d, the place.
8	15	16	Have you had soundings, and how many fathoms had you? *Ans.* From the Compound Table, or General Answers, 17th Chap.
9	17	18	What bottom had you?
10	19	20	Is there a current here, and which way does it set?
11	21	22	How many knots does the current run? *Ans.* From the Compound Table, with number.
12	23	24	Which passage do you intend to take? *Ans.* From Compound Table, name of the place.
13	25	26	I wish to try rate of sailing.
14	27	28	What course by compass do you steer? *Ans.* From Compound Table.
15	29	30	Shall we steer for the *place*, named by the next pair of signals (per Compound Table)?
16	31	32	What course do you intend to steer? *Ans.* From Compound Table.
17	33	34	What course do you intend to steer to-night? *Ans.* From Compound Table.

Numerical order of Sentences.	Pair Signals. No.	No.	CHAPTER I. *Relating to the Navigation of Ships.*
			QUESTIONS.
18	35	36	What course does she, or do they steer? *Ans.* From Compound Table.
19	37	38	At what hour have we high water? *Ans.* From Compound Table.
20	39	40	At what hour have we low water? *Ans.* From Compound Table.
21	2	1	What tide have we now?
22	4	3	Do you think I shall be able to beat up?
23	6	5	What advice can you give me for beating up here?
24	8	7	Do you think it safe to continue our present course, and how long?
			General Note.—In regard to those sentences, where the Compound Table is not referred to for an answer, the appropriate answer will be found under the second subdivision of the same chapter, or among the general answers in the 17th chapter.
			ANSWERS OR COMMUNICATIONS.
25	10	9	I have no chronometer.
26	12	11	My *latitude* by dead reckoning is as follows:—(per Compound Table.)
27	14	13	My *latitude* by double altitudes is as follows:—(per Compound Table.)
28	16	15	My *latitude* from an altitude of the polar star is as follows: (per Compound Table.)
29	18	17	My *latitude* by observation is as follows:—(per Compound Table).
30	20	19	My *longitude* by lunar observation is as follows:—(per Compound Table).
31	22	21	My *longitude*, lately corrected from bearings of the land, is as follows:—(per Compound Table.)
32	24	23	My *longitude*, as I had it from another ship, is as follows:—(per Compound Table.)
33	26	25	My *longitude* by chronometer is as follows:—(per Compound Table.)
34	28	27	No bottom.
35	30	29	Rocky bottom.
36	32	31	Sand and shells.
37	34	33	Sand.

Numerical order of Sentences.	Pair Signals. No.	No.	CHAPTER I. *Relating to the Navigation of Ships.* ANSWERS OR COMMUNICATIONS.
38	36	35	Coarse Sand.
39	38	37	Fine Sand.
40	40	39	Pebbles or stony bottom.
			Note.—In using the following sentences it will be necessary to change the flags, the second signal being composed of flags different or differently arranged from those of the first.
41	1	3	Sand with red spots.
42	1	4	Sand with black spots.
43	1	5	Clay.
44	1	6	Muddy bottom.
45	1	7	No current.
46	1	8	Current sets on shore.
47	1	9	Current sets off shore.
48	1	10	Current is with us.
49	1	11	Current is against us.
50	1	12	Flood tide is setting now.
51	1	13	Ebb tide is setting now.
52	1	14	It is slack water now.
53	1	15	I'll try rate of sailing with you
54	1	16	I cannot try rate of sailing with you.
55	1	17	I shall continue during the night the course I am steering now.
56	1	18	I shall alter my course to *the point of compass,* and at *the hour* indicated, by the two following pair of signals :—(per Compound Table.)
57	1	19	I shall stand off, to *the hour,* indicated by the pair of signals following:—(per Compound Table.)
58	1	20	I shall steer during the night *the course,* as indicated by the next pair of signals :—(per Compound Table.)
59	1	21	I shall steer to *the hour, the course* indicated by the two following pair of signals:—(per Compound Table.)
60	1	22	I shall, from the last-mentioned hour, alter my course to the *point of compass* indicated by the pair of signals following:—(per Compound Table.)
61	1	23	I shall not alter my course until I get sight of the land.
62	1	24	If you think of beating up, you must keep the northern shore.
63	1	25	If you think of beating up, you must keep the eastern shore.
64	1	26	If you think of beating up, you must keep the southern shore.
65	1	27	If you think of beating up, you must keep the western shore.
66	1	28	If you think of beating up, you must make short tacks.
67	1	29	If you think of beating up, you must not keep the shore.

Numerical order of Sentences.	Pair Signals. No.	No.	CHAPTER I. *Relating to the Navigation of Ships.* ANSWERS OR COMMUNICATIONS.
68	1	30	If you think of beating up, you must keep near the shore, or near the *land*, indicated by the signal following :—(per Compound Table.) *Note.*—The Shore or Coast may be indicated by a place situated on it. If the land, or coast in question, has been previously mentioned, or cannot otherwise be mistaken, no second pair of signals is needed ; and, to indicate this, the last signal, when answered, is hauled nearly down, hoisted a second time, and then finally struck.
	1	31	If you continue on your present course, you will run into danger.
70	1	32	The course may be continued to the *hour* indicated by the pair of signals following :—(per Compound Table.)

CHAPTER II.

Numerical order of Sentences.	Pair Signals.		Relating to Land and Coasts; Channels and Entrances; Harbours and Anchoring-places; Lights, and other Sea-marks; Pilots, Shore-going, and Watering.
	No.	No.	QUESTIONS.
71	1	33	Have you seen the land?
72	1	34	Is it advisable to stand nearer in?
73	1	35	How near may the land be approached with safety (answered by number of leagues from Compound Table), with fractions or without, or fractions alone.—See Introduction about Fractions.
74	1	36	How near may the land be approached, at the *place* indicated by the following pair of signals (per Compound Table)? *Ans.* From Compound Table, as above.
75	1	37	How many leagues do you reckon yourself to be from the *place* indicated by the following pair of signals (per Compound Table)? *Ans.* From Compound Table.
76	1	38	How does the land bear? *Ans.* From Compound Table, points of the compass.
77	1	39	How did the land bear when you saw it last? *Ans.* From Compound Table.
78	1	40	In what point of the compass had you the *place* indicated by the following pair of signals (per Compound Table), when you last saw it? *Ans.* From Compound Table.
79	2	3	What land do you make that in sight to be? *Ans.* From Compound Table, or from 17th Chapter
80	2	4	What land did you make that to be, which we had a glimpse of just now? *Ans.* From Compound Table, or 17th Chapter.
81	2	5	Shall you try to make the land to-night?
82	2	6	Are you acquainted with the coast? *Ans.* Yes, or No; or find an appropriate reply under the answering division of this Chapter.
83	2	7	Are you acquainted with the channel or entrance?
84	2	8	Is the vessel, lying in the *direction* from me to be indicated by the following pair of signals (per Compound Table), acquainted with the channel or entrance?

CHAPTER II.

Relating to Land and Coasts; Channels and Entrances; Harbours and Anchoring-places; Lights; Pilots, &c.

Numerical order of Sentences.	Pair Signals. No.	No.	QUESTIONS.
85	2	9	How does the channel or entrance bear?
			Ans. From Compound Table.
86	2	10	Will you lead through the channel or entrance?
87	2	11	How does the harbour bear of us?
			Ans. From Compound Table.
88	2	12	Is there water enough for your ship to enter the harbour in sight, or in question, or *to be* indicated by the following pair of signals (per Compound Table)?
			Ans. From 17th Chapter.
			Note.—If the harbour is in sight, or has been named in the foregoing conversation, the last signal of the pair, when answered, is hauled nearly down, then hoisted aloft, and finally struck, to indicate that no signal for the name of the harbour is to follow.
89	2	13	With what draught of water may the harbour in sight, or in question, or *to be* indicated by the signal following (per Compound Table) be entered? (See the note above.)
			Ans. By number of feet, from Compound Table.
90	2	14	What is the name of that harbour?
			Ans. From Compound Table.
91	2	15	Have you ever been in that harbour?
92	2	16	I will follow you, if you are acquainted with the harbour?
93	2	17	Are you going into port?
94	2	18	What harbour do you intend to make?
			Ans. From Compound Table.
95	2	19	Do you intend to run in to-night?
96	2	20	Shall we make a harbour?
97	2	21	May we expect many vessels from that harbour?
98	2	22	Is there any shelter?
99	2	23	Have you any knowledge of the anchorage?
100	2	24	Is the anchorage safe with all winds?
101	2	25	What anchorage have you obtained?
102	2	26	I shall follow you, if you'll lead me to the best berth.
103	2	27	Am I in a good berth now?
104	2	28	What wind must we have to get in?
			Ans. From Compound Table, points of the compass.
105	2	29	How is the road to lie in?
106	2	30	What bottom for anchorage is there?
107	2	31	On what depth is the best anchorage?
			Ans. By numbers of feet from Compound Table.

CHAPTER II.

Relating to Land and Coasts; Channels and Entrances; Harbours and Anchoring-places; Lights; Pilots, &c.

Numerical order of Sentences.	Pair Signals. No.	No	QUESTIONS.
108	2	32	On what side of the road, harbour, or bay, is the best anchorage?
			Ans. With point of compass from Compound Table.
109	2	33	May we run in by the lights to-night?
110	2	34	Shall you keep under the lights to-night?
111	2	35	Is a pilot to be had here?
112	2	36	Have you a pilot on board?
113	2	37	Shall I send you a pilot?
114	2	38	Do you take a pilot here?
115	2	39	If you are going ashore, I'll thank you for a passage in your boat?
116	2	40	Are you going on shore?
117	3	1	If you send a boat on shore, will you take a message for me?
118	3	2	May we land in our own boats?
119	3	5	Have you had a boat on shore?
120	3	6	Have you been on shore?
121	3	7	Have you had any communication with the shore?
122	3	8	Shall we meet on shore?
123	3	9	At what hour do you come on board?
			Ans. From Compound Table, or 17th Chapter.
124	3	10	Do you remain on shore to-night?
			Ans. By Yes—or No.
1 5	3	11	Have you settled all your affairs on shore?
126	3	12	Is there a convenient place here for filling our water-casks?
127	3	13	May we have water sent off to us?
128	3	14	Is there a convenient watering-place at the *place* indicated by the following pair of signals?—(per Compound Table.)
129	3	15	Must you water here?
130	3	16	How much water have you on board?
			Ans. From Compound Table, by number of gallons.
			ANSWERS OR COMMUNICATIONS.
131	3	17	Keep a good look out for the land.
132	3	18	Land is in sight.
133	3	19	I have seen the land.
134	3	20	I shall try to make the land to-night.
135	3	21	I have discovered land in the *point of the compass*, indicated by the following pair of signals:—(per Compound Table.)

Numerical order of Sentences.	Pair Signals. No.	No.	CHAPTER II. *Relating to Land and Coasts; Channels and Entrances; Harbours and Anchoring-places; Lights; Pilots, &c.* ANSWERS OR COMMUNICATIONS.
136	3	22	I shall not try to make the land to-night, but heave to, or lie by, from the *hour* indicated by the following pair of signals:—(per Compound Table.)
137	3	23	I shall not make the land to-night, but stand off from the *hour* indicated by the following pair of signals:—(per Compound Table.)
138	3	24	You are too near the land.
139	3	25	Keep more off shore.
140	3	26	Steer out from the land.
141	3	27	I shall stand off all night.
142	3	28	I shall stand off till the *hour* indicated by the following pair of signals:—(per Compound Table.)
143	3	29	I shall stand in for the land as long as I can see.
144	3	30	She is standing off shore.
145	3	31	She is standing in for the land.
146	3	32	You ought not to come nearer the land than the *distance*, in leagues, indicated by signal following:—(either by numbers from Compound Table, or by fractions, if less than a league, or by both, if required.)
147	3	33	Coast is dangerous.
148	3	34	Coast is not dangerous.
149	3	35	I have discovered an entrance or inlet, in the *point of compass*, as indicated by the following pair of signals:—(per Compound Table.)
150	3	36	I shall lead the way through the channel or entrance.
151	3	37	I am going into port.
152	3	38	Vessel or boat in the *direction* from me, as indicated by the following pair of signals (per Compound Table), is steering for the harbour.
153	3	39	The harbour is right a-head.
154	3	40	We are to leeward of the harbour.
155	4	1	We are to windward of the harbour.
156	4	2	Harbour *bears* of me, as indicated by the following pair of signals:—(per Compound Table.)
157	4	5	Harbour is good.
158	4	6	Harbour is bad.
159	4	7	Harbour is bad, with the wind from the *point of compass*, indicated next (per Compound Table).
160	4	8	I am acquainted with the harbour, you may steer after me.
161	4	9	I am not acquainted with the harbour.
162	4	10	Harbour is a bar-harbour.
163	4	11	Anchoring-place is good with all winds.

Numerical order of Sentences.	Pair Signals.		CHAPTER II. *Relating to Land and Coasts; Channels and Entrances; Harbours and Anchoring-places; Lights; Pilots, &c.*
	No.	No.	ANSWERS OR COMMUNICATIONS.
164	4	12	Anchoring-place is good with the wind that blows now.
165	4	13	Anchoring-place is safest with the wind from the *point of compass* indicated next (per Compound Table).
166	4	14	If you will follow me, I will lead you to the best berth.
167	4	15	The best berth is in the *point of compass* indicated next (per Compound Table).
168	4	16	I have obtained anchorage in the *number* of fathoms indicated next (per Compound Table).
169	4	17	Good bottom for anchorage.
170	4	18	Bad bottom for anchorage.
171	4	19	You are in a good berth now.
172	4	20	You are not in a good berth.
173	4	21	I shall keep under the light (keep the light in sight) all night.
174	4	22	I shall run in to-night by the lights.
175	4	23	Keep under the light to-night.
176	4	24	Floating light, at the *place* indicated next (per Compound Table) has parted her moorings.
177	4	25	Floating light, which had been driven from her moorings, is returned to her *place*, the name of which will be indicated next (per Compound Table).
178	4	26	A floating light is laid out at the *place* indicated next (per Compound Table).
179	4	27	The buoy or floating mark at the *place* indicated next (per Compound Table) has been taken up, or has drifted away.
180	4	28	The buoy, or floating mark, at the *place* indicated next (per Compound Table) is not to be relied on after the last storm, or ice.
181	4	29	The buoys, or marks, at the *place* indicated next (per Compound Table) have drifted away, or have been taken up.
182	4	30	The buoys, or marks, at the *place* indicated next (per Compound Table) are not to be relied on after the late gale, or ice.
183	4	31	The northerly buoy or mark at the *place* indicated next (per Compound Table), is not in its right place.
184	4	32	The eastern buoy or mark, at the *place* indicated next (per Compound Table), is not in its right place.
185	4	33	The southern buoy or mark at the *place* indicated next (per Compound Table), is not in its right place.
186	4	34	The westerly buoy or mark at the *place* indicated next (per Compound Table), is not in its right place.
187	4	35	Send a pilot on board.
188	4	36	If you come on shore before me, send a pilot off.

Numerical order of Sentences.	Pair Signals. No.	No.	CHAPTER II. *Relating to Land and Coasts; Channels and Entrances; Harbours and Anchoring-places; Lights; Pilots, &c.* ANSWERS OR COMMUNICATIONS.
189	4	37	You had better take a pilot.
190	4	38	There are no pilots to be had.
191	4	39	You need not take a pilot, but may at all events steer after me.
192	4	40	I am going on shore.
193	5	1	I shall meet you on shore.
194	5	2	I am not going on shore to-day.
195	5	3	I have not been on shore to-day.
196	5	4	I have not been on shore since my arrival.
197	5	7	I shall not go on shore here.
198	5	8	I am going on shore, and will give you a passage.
199	5	9	I shall not go on shore till the *hour* indicated next (per Compound Table).
200	5	10	I shall send you my boat before it goes on shore.
201	5	11	I am going to send a boat on shore.
202	5	12	I shall not send on shore.
203	5	13	Boat is a-shore.
204	5	14	Mate is a-shore.
205	5	15	I have had no communication with the shore.
206	5	16	I shall have no communication with the shore.
207	5	17	You'll do better to have no communication with the shore.
208	5	18	It is difficult to land here.
209	5	19	There is too much surf to land.
210	5	20	Surf is not so great as to prevent landing.
211	5	21	I shall be obliged to water.
212	5	22	You'll have to water here.
213	5	23	There is good watering here.
214	5	24	Place is not good for watering.
215	5	25	Watering must be done with own men and boat.
216	5	26	Watering cannot well be done, but with men and boat from the shore.

CHAPTER III.

Numerical order of Sentences.	Pair Signals. No.	Pair Signals. No.	*Relating to Ship Manœuvres.*
			QUESTIONS.
217	5	27	At what time do you intend to weigh?
218	5	28	Shall you anchor?
219	5	29	Will you keep close to me during the night?
220	5	30	Shall I carry a light during the night, for you to follow me?
221	5	31	Shall you heave to?
222	5	32	Shall we heave to?
223	5	33	If you will heave to, I will send my boat on board.
224	5	34	I shall heave to, if you'll send your boat on board.
225	5	35	Shall you moor?
226	5	36	At what hour will you tack or wear?
			Ans. From Compound Table.
227	5	37	What is the reason you are keeping away so much?
228	5	38	What is the reason you are luffing up so much?
229	5	39	Shall we haul our wind?
230	5	40	Shall we bear up?
231	6	1	Shall we keep under easy sail?
232	6	2	Shall we carry more sail?
233	6	3	Do you think we may ride it out, or should we cut and run?
			ANSWERS OR COMMUNICATIONS.
234	6	4	I shall weigh immediately.
235	6	7	I shall weigh as soon as the wind abates.
236	6	8	I shall weigh as soon as there is a breeze.
237	6	9	I shall not weigh till I have a fair wind.
238	6	10	I shall weigh upon the ebb.
239	6	11	I shall weigh upon the flood.
240	6	12	I shall weigh at slack water.
241	6	13	I shall weigh, wind and weather permitting.
242	6	14	I shall weigh to-day.
243	6	15	I shall weigh at the *hour* indicated next (per Compound Table).
244	6	16	I shall weigh to-night.
245	6	17	I shall weigh to-morrow morning.
246	6	18	I shall weigh whenever I can.
247	6	19	I shall weigh as soon as my boat returns.
248	6	20	You must look to and get under weigh, that I may likewise go about it.

Numerical order of Sentences.	Pair Signals. No.	No	CHAPTER III. *Relating to Ship Manœuvres.* ANSWERS OR COMMUNICATIONS.
249	6	21	Carry a light during the night.
250	6	22	I shall carry a light during the night.
251	6	23	Show a light when you go about (tack or wear) to-night.
252	6	24	I shall hoist a light when I tack or wear to-night.
253	6	25	I shall follow you, if you will carry a light to-night.
254	6	26	Make sail a-head.
255	6	27	Keep on starboard bow.
256	6	28	Keep on starboard beam.
257	6	29	Keep abaft beam on starboard.
258	6	30	Keep before beam on larboard.
259	6	31	Keep on larboard beam.
260	6	32	Keep abaft larboard beam.
261	6	33	Keep astern.
262	6	34	Keep on the other tack.
263	6	35	Keep on this tack.
264	6	36	Do not stand too far on your present tack.
265	6	37	I shall stand on this tack until the *hour* indicated next (per Compound Table).
266	6	38	I shall stand on the other tack until the *hour* indicated next (per Compound Table).
267	6	39	We shall do better on the other tack.
268	6	40	Keep away, or keep more away.
269	7	1	I shall keep away.
270	7	2	Haul your wind, or haul nearer to the wind.
271	7	3	Haul your wind on starboard tack.
272	7	4	Haul your wind on larboard tack.
273	7	5	I shall haul to the wind, or come nearer to the wind.
274	7	6	I shall keep close hauled all night.
275	7	9	Vessel is bearing up, or has borne up.
276	7	10	Vessel is hauling her wind, or has hauled to the wind.
277	7	11	Tack if you can.
278	7	12	I shall tack about.
279	7	13	Vessel has tacked, or gone about.
280	7	14	I shall wear round.
281	7	15	Keep within sight.
282	7	16	Keep me in sight during the night.
283	7	17	I shall keep close to you during the night.
284	7	18	I shall stand off and on.
285	7	19	I shall stand off and on all night.
286	7	20	I shall close reef.
287	7	21	I shall reef courses.
288	7	22	I shall keep under easy sail during the night.

CHAPTER III.

Relating to Ship Manœuvres.

Numerical order of Sentences.	Pair Signals. No.	No.	ANSWERS OR COMMUNICATIONS.
289	7	23	Keep under easy sail all night.
290	7	24	I mean to carry on all night, as I am anxious to get out.
291	7	25	I shall carry all sail possible.
292	7	26	Make sail, or make more sail.
293	7	27	You must make more sail, if you wish to keep company.
294	7	28	I cannot make more sail, as my masts and rigging can carry no more.
295	7	29	Shorten sail.
296	7	30	You must shorten sail, as I cannot keep up.
297	7	31	Shorten sail, before the squall comes on.
298	7	32	Vessel is on the starboard tack.
299	7	33	Vessel is on the larboard tack.
300	7	34	Vessel is close hauled.
301	7	35	Vessel is going free, or before the wind.
302	7	36	Vessel seen has all sail set.
303	7	37	I shall anchor.
304	7	38	I shall anchor, if circumstances will permit.
305	7	39	I shall anchor, rather than be carried back by the tide or current.
306	7	40	I must come to an anchor, for fear of being set on shore by the current.
307	8	1	Water is too deep to anchor.
308	8	2	Do not anchor, as there is danger.
309	8	3	I shall heave to.
310	8	4	I shall heave to all night.
311	8	5	I shall heave to, until you are out.
312	8	6	I dare not heave to here.
313	8	9	I shall heave to, from dark till the *hour* indicated next (per Compound Table).
314	8	10	I shall moor.
315	8	11	I shall not moor.
316	8	12	I shall make fast by hawsers on shore.
317	8	13	It is not safe here, unless you moor.
318	8	14	You will be obliged to bring out a kedge, to keep your anchor clear.
319	8	15	Veer cable, or veer more cable.
320	8	16	Shorten cable, or shorten more in upon your cable.
321	8	17	I cannot veer more cable.
322	8	18	I cannot shorten more in upon my cable.
323	8	19	I am hove short.
324	8	20	I cannot purchase my anchor.
325	8	21	Cut or slip your cable, and make all sail possible to get out.
326	8	22	I cannot wait any longer for you.

CHAPTER IV.

Numerical order of Sentences.	Pair Signals. No.	No	*Relating to Wind and Weather.* QUESTIONS.
327	8	23	What do you think of the weather?
328	8	24	Do you think the wind will continue as it is? or, do you think we shall soon have a change of wind?
329	8	25	What does your barometer tell?
330	8	26	What wind have you had?
331	8	27	Have you had variable winds?
332	8	28	Has vessel belonging to the *place* indicated next (per Compound Table), and whose name shall be spelled, or vessel aforementioned, had variable winds? *Note.*—If the vessel has been mentioned in the foregoing conversation, so that no further mentioning of her name, or the place she belongs to, is necessary, the last signal of the pair, when answered, is hauled nearly down, then rehoisted, and struck again.
333	8	29	How has the weather been at the place you come from?
334	8	30	Has there been a storm or hurricane at the place you come from?
335	8	31	Has there been a storm or hurricane at the *place* indicated next (per Compound Table), and has it done much damage?
336	8	32	Has there been much rain at the place you come from, or *as* indicated next (per Compound Table)? *Note.*—If the question refers to the place the other is coming from, the last signal of the pair, when answered, is hauled nearly down, rehoisted and struck.
337	8	33	Is the winter set in at the place you are coming from, or *which* will be indicated next (per Compound Table)? *Note.*—See the preceding note.
338	8	34	Is the *bay, firth, or harbour,* indicated next (per Compound Table) frozen over?
339	8	35	Is there a possibility of getting through the ice at the *place* indicated next (per Compound Table)? To be answered by Yes, or No.
340	8	36	Where did you fall in with the ice? *Ans.* From Compound Table, Latitude and Longitude.
341	8	37	Is the ice dangerous? *Ans.* Yes, or No.
342	8	38	Where do you think I may get into to avoid the ice? *Ans.* Place from Compound Table.

Numerical order of Sentences.	Pair Signals. No.	No.	CHAPTER IV. *Relating to Wind and Weather.* QUESTIONS.
343	8	39	Have you met floating ice? *Ans.* Yes, or No.
344	8	40	Had there been a plentiful fall of snow at the place you come from, or *as* indicated next (per Compound Table)? *Note.*—If the place the other is coming from is meant, the last signal of the pair, when answered, is hauled nearly down, rehoisted, and then struck.
345	9	1	Is the ice broken up? *Ans.* Yes, or No.
346	9	2	Is the sea clear from ice?
347	9	3	Has the N.E. monsoon set in, and in what latitude and longitude did you fall in with it? *Ans.* No, or by latitude and longitude from Compound Table.
348	9	4	Has the S.W. monsoon set in, and in what latitude and longitude did you fall in with it? *Ans.* No, or by latitude and longitude from Compound Table.
349	9	5	Has the N.W. monsoon set in, and in what latitude and longitude did you fall in with it? *Ans.* No, or by latitude and longitude from Compound Table.
350	9	6	Has the S.E. monsoon set in, and in what latitude and longitude did you fall in with it? *Ans.* No, or by latitude and longitude from Compound Table.
			ANSWERS OR COMMUNICATIONS.
351	9	7	It looks like wind.
352	9	8	It looks like rain.
353	9	11	It looks like snow.
354	9	12	I think we shall have thunder.
355	9	13	I think a fog is coming on.
356	9	14	It looks like a heavy squall coming on.
357	9	15	It looks like a hurricane coming on.
358	9	16	I think we shall have no hurricane.
359	9	17	I think there will be no change of weather.
360	9	18	The breeze freshens.
361	9	19	The breeze is dying away.
362	9	20	It looks like falling calm.
363	9	21	It is dangerous to be becalmed here.
364	9	22	I have no barometer.
365	9	23	The barometer is rising.

Numerical order of Sentences.	Pair Signals. No.	No.	CHAPTER IV. *Relating to Wind and Weather.* ANSWERS OR COMMUNICATIONS.
366	9	24	The barometer is falling.
367	9	25	The barometer stands above variable.
368	9	26	The barometer stands below variable.
369	9	27	The barometer foretels a storm.
370	9	28	The barometer is stationary.
371	9	29	The barometer indicates good weather.
372	9	30	The barometer seems to be rising.
373	9	31	The barometer seems to be falling.
374	9	32	I think we may expect the wind from the *point of compass* indicated next (per Compound Table).
375	9	33	I think the wind will continue as we have it now.
376	9	34	I have had variable winds.
377	9	35	I have had the wind, for the most part, from the *point of compass* indicated next (per Compound Table).
378	9	36	I have had the wind fair.
379	9	37	I have had the wind fair, but calm.
380	9	38	Vessel spoken of has had baffling winds.
381	9	39	Vessel spoken of has had the wind, for the most part, from quarter indicated (per Compound Table, point of compass).
382	9	40	The prevailing wind at the place I am coming from, has been from quarter indicated next (per Compound Table, point of compass).
383	10	1	I have had a hard gale from the quarter indicated next (per Compound Table, point of compass).
384	10	2	Weather has been mostly calm at the place I have left.
385	10	3	I have had a hurricane on the *date*, and in *latitude*, that will be indicated by the two following pairs of signals:—(per Compound Table, and in the same order as mentioned here.)
386	10	4	There has been a hurricane at the place I have left.
387	10	5	There has been a hurricane on the *date*, and at the *place*, that will be indicated by the two pairs of signals following:—(per Compound Table, and in the same order as mentioned here.)
388	10	6	The wind has been varying between the two *points of compass*, indicated by the following two pairs of signals:—(per Compound Table).
389	10	7	The wind has been pretty steady from quarter indicated next, (per Compound Table, point of compass).
390	10	8	The hurricane, or gale, has occasioned no damage among the shipping.

Numerical order of Sentences.	Pair Signals.		CHAPTER IV. *Relating to Wind and Weather.*
	No.	No.	ANSWERS OR COMMUNICATIONS.
391	10	11	The hurricane, or gale, has done damage to the shipping. *Note.*—For average, see Chap. VII.
392	10	12	The hurricane, or gale, has done no damage on shore.
393	10	13	The hurricane, or gale, has done damage on shore.
394	10	14	The hurricane raged between the *islands*, or *places*, indicated by the two following pairs of signals:—(per Compound Table.)
395	10	15	There has been no hurricane yet this year.
396	10	16	There has been a long drought at the place I come from.
397	10	17	Weather has been variable, with rain and dry weather, at the place I come from.
398	10	18	There has been a good season of rain at the place I come from.
399	10	19	The winter had not set in at the place I came from, or which is spoken of.
400	10	20	The sea was already frozen over, at the place I come from, or which is spoken of.
401	10	21	The *place* indicated next (per Compound Table) is locked up by the ice.
402	10	22	You will hardly be able to get through the ice, at the *place* indicated next (per Compound Table).
403	10	23	The ice is broken up at the place I come from, or *which* will be indicated next (per Compound Table). *Note:*—If the place the vessel is coming from is meant, the last signal when answered, is hauled nearly down, rehoisted, and then struck.
404	10	24	The ice is not broken up yet, at the place spoken of, or *which* will be indicated next (per Compound Table). *Note.*—See the note to sentence preceding.
405	10	25	I see ice or icebergs, in the *point of compass* from me, as indicated next (per Compound Table).

CHAPTER V.

Numerical order of Sentences.	Pair Signals. No.	No.	*Relating to Quarantine, state of Health, Sickness, &c.* QUESTIONS.
406	10	26	Are you all well ?
407	10	27	How does the person do, whose name I am going to spell ?
408	10	28	Are all well at home ?
409	10	29	Are my family well ?
410	10	30	Have you heard from home?
411	10	31	Does any contagious disease exist at the place you come from, or *which* will be indicated next (per Compound Table)?
			Ans. By No, or look for answer.
			Note.—If the place is understood from which the vessel is coming, the last signal, when answered, is hauled nearly down, rehoisted, and then struck.
412	10	32	Have you heard any thing of the person whose name I am going to spell ?
413	10	33	Has the contagious disease, or the plague, ceased at the place you come from ?
414	10	34	Were all well on board the vessel, belonging to *place* indicated next (per Compound Table) and whose name will be spelt ?
415	10	35	Has the contagious disease, or the plague, ceased at the *place* indicated next (per Compound Table) ?
416	10	36	Have you many sick on board ?
			Ans. With a *cipher* or *number*, from Compound Table.
417	10	37	Is the sickness contagious ?
418	10	38	What is the sickness you have on board ?
419	10	39	Have you a doctor ?
420	10	40	Will you allow your doctor to come on board ?
421	11	1	Will you send for the doctor ?
422	11	2	Will you send me a doctor from shore ?
423	11	3	What is the name of the passenger sick on board ?
			Ans. By spelling.
424	11	4	What is the name of the passenger dead on board ?
			Ans. By spelling the name.
			Note.—Should there be more than one sick, or dead, the name of each person may be given in succession by spelling.
425	11	5	Shall I be put under quarantine at the place where I am going

Numerical order of Sentences.	Pair Signals. No	No.	CHAPTER V. *Relating to Quarantine, state of Health, Sickness, &c.*
			QUESTIONS.
			to, or *which* will be indicated next (per Compound Table). *Note.*—If the place which the vessel is bound to is meant, and has been named before, the last signal, when answered, is hauled nearly down, then rehoisted, and struck.
426	11	6	How many days quarantine shall I be put under? *Ans.* From Compound Table, by number.
427	11	7	What sickness is prevailing at the place you come from, or *which* will be indicated next (per Compound Table)? *Note.*—If the place which the vessel is coming from is understood, the last signal when answered, must be hauled nearly down then rehoisted, and finally struck.
			ANSWERS OR COMMUNICATIONS.
428	11	8	All are well.
429	11	9	Captain is well.
430	11	10	Captain is unwell, or sick.
431	11	13	Captain is dead.
432	11	14	I have a sick passenger.
433	11	15	Several of my passengers are sick.
434	11	16	A passenger is dead.
435	11	17	Several of my passengers are dead; *number* will be indicated by next signal (per Compound Table).
436	11	18	The person you are making inquiry for is well.
437	11	19	The person you are making inquiry about is unwell or sick.
438	11	20	The person you are making inquiry about is dead.
439	11	21	The sickness on board is not contagious.
440	11	22	I have contagious sickness on board.
441	11	23	I have the yellow fever on board.
442	11	24	I have the Asiatic cholera on board.
443	11	25	I have the smallpox on board.
444	11	26	I have the plague on board.
445	11	27	Your family are all well.
446	11	28	Your wife has presented you with a son.
447	11	29	Your wife has presented you with a daughter.
448	11	30	Your wife has presented you with twins.
449	11	31	The yellow fever has broke out at the place you were inquiring about, or *which* will be indicated next (per Compound Table). *Note.*—When the place inquired about is understood, the last signal, when answered, must be hauled nearly down, then rehoisted, and finally struck.

Numerical order of Sentences.	Pair Signals. No.	No.	CHAPTER V. Relating to Quarantine, state of Health, Sickness, &c. — ANSWERS OR COMMUNICATIONS.
450	11	32	The Asiatic cholera has broken out at the place you are inquiring about, or *which* will be indicated next (per Compound Table). *Note.*—What is said in the note to the foregoing sentence applies equally to this.
451	11	33	The plague has broken out at the place you are inquiring about, or *which* will be indicated next (per Compound Table). *Note.*—What is said in the note to the preceding sentences is equally applicable to this.
452	11	34	The smallpox has broken out at the place you are inquiring about, or *which* will be indicated next (per Compound Table). *Note.*—See notes to preceding sentences.
453	11	35	Contagious disease is reigning at the place you are inquiring about, or *which* will be indicated next (per Compound Table). *Note.*—See notes to preceding sentences.
454	11	36	There is no contagious disease at the place I come from.
455	11	37	There is yellow fever at the place I come from.
456	11	38	There is Asiatic cholera at the place I come from.
457	11	39	There is plague at the place I come from.
458	11	40	There is smallpox at the place I come from.
459	12	1	There is contagious disease at the place I come from.
460	12	2	Vessel belonging to the *place* indicated next (per Compound Table), and whose name shall be spelt, has the yellow fever on board.
461	12	3	Vessel belonging to the *place* indicated next (per Compound Table), and whose name shall be spelt, has the Asiatic cholera on board.
462	12	4	Vessel belonging to the *place* indicated next (per Compound Table), and whose name shall be spelt, has the plague on board.
463	12	5	Vessel belonging to the *place* indicated next (per Compound Table), and whose name shall be spelt, has the smallpox on board.
464	12	6	Vessel belonging to the *place* indicated next (per Compound Table), and whose name shall be spelt, has contagious disease on board.
465	12	7	You will be put under quarantine.
466	12	8	Vessel belonging to the *place* indicated next (per Com-

Numerical order of Sentences.	Pair Signals. No.	No.	CHAPTER V. *Relating to Quarantine, state of Health, Sickness, &c.* ANSWERS OR COMMUNICATIONS.
			pound Table), and whose name shall be spelt, is under quarantine. *Note.*—This sentence may likewise apply to vessel in question, or already spoken of; and in such case, the last signal, when answered, must be hauled nearly down, then rehoisted, and finally struck.
467	12	9	Vessel belonging to the *place* indicated next (per Compound Table), and whose name shall be spelt, is lying under quarantine at the *place*, which will be indicated by the second pair of signals following:—(from Compound Table.) *Note.*—If the signal alludes to a vessel already spoken of, or named in the foregoing conversation, the last signal of this sentence, when answered, is hauled nearly down, then rehoisted, and finally struck; whereby it is indicated that no signals for the place the vessel belongs to, and no spelling of her name are to follow.
468	12	10	I shall send you my doctor.
469	12	13	My doctor is sick.
470	12	14	I have no doctor.
471	12	15	My doctor is dead.

CHAPTER VI.

Numerical order of Sentences.	Pair Signals. No.	No.	*Relating to Packets, Steam-boats, Mails, Letters, and Newspapers.* QUESTIONS.
472	12	16	Has the packet from the *place* indicated next (per Compound Table) arrived? *Note.*—When only one packet is plying, no signal indicating place need be made, and in this case the last signal, when answered, is hauled nearly down, rehoisted, and struck.
473	12	17	At what date did the packet from the *place* indicated next (per Compound Table) arrive? *Ans.* From Compound Table. *Note.*—See note to the preceding sentence.
474	12	18	Has the packet sailed?
475	12	19	Has the packet for the *place* indicated next (per Compound Table) sailed?
476	12	20	What date did the packet sail? *Ans.* From Compound Table.
477	12	21	What date did the packet sail for the *place* indicated next (per Compound Table)? *Ans.* From Compound Table.
478	12	22	On what day, or days, of the week, do the sailing-packets start for the place we have been speaking of, or *which* will be indicated next (per Compound Table)? *Ans.* From Compound Table, days of the week. *Note.*—If the place spoken of is to be understood, the last signal, when answered, is hauled nearly down, rehoisted, and struck.
479	12	23	Has the steam-vessel arrived from the *place* indicated next (per Compound Table)?
480	12	24	At what date did the steam-vessel arrive? *Ans.* From Compound Table.
481	12	25	At what date did the steam-vessel arrive from the *place* indicated next (per Compound Table)? *Ans.* From Compound Table.
482	12	26	Has the steam-vessel sailed?
483	12	27	Has the steam-vessel for the *place* indicated next (per Compound Table) sailed?
484	12	28	At what date did steam-vessel sail? *Ans.* From Compound Table.

Numerical order of Sentences.	Pair Signals. No.	No	CHAPTER VI. *Relating to Packets, Steam-boats, Mails, Letters, and Newspapers.* QUESTIONS.
485	12	29	At what date did the steam-vessel sail for the *place* indicated next (per Compound Table)? *Ans.* From Compound Table.
486	12	30	What opportunity by water is there from this to the *place* indicated next (per Compound Table)?
487	12	31	What opportunity by water is there between the two *places* indicatedbythefollowing two pairs of si gnals?—(per Compound Table.)
488	12	32	What hour does the steam-vessel sail from this to the *place* indicated next (per Compound Table)? *Ans.* From Compound Table.
489	12	33	On what day or days of the week does the steamer sail from this to the place spoken of, or *which* will be indicated next (per Compound Table)? *Ans.* From Compound Table, day or days of the week. *Note.*—If the place need not be named, as being already mentioned in the foregoing conversation, the last signal, when answered, is hauled nearly down, then rehoisted, and struck.
490	12	34	Is there no other opportunity by water from this place than the steam-vessel?
491	12	35	Is the mail arrived?
492	12	36	Is the mail gone?
493	12	37	Is the mail from the *place* indicated next (per Compound Table) arrived? *Note.*—See note to sentence following.
494	12	38	Is the mail to the *place* indicated next (per Compound Table) gone? *Note.*—At places where only one mail exists, some of the colonies and islands for example, no signal need be made for the place the mail is going to, and in such case the last signal—in this and preceding sentence—when answered, is hauled nearly down, then rehoisted, and struck.
495	12	39	What hour is the latest for delivering letters at the post-office? *Ans.* From Compound Table.
496	12	40	What is the date of your latest letter from home, or from the *place* indicated next (per Compound Table)? *Ans.* From Compound Table. *Note.*—If "from home" is to be understood, the last signal.
497	13	1	Have you any letters for me?
498	13	2	Have you any letters for the person whose name I am going to spell?

CHAPTER VI.

Relating to Packets, Steam-boats, Mails, Letters, and Newspapers.

Numerical order of Sentences.	Pair Signals. No.	No.	QUESTIONS.
499	13	3	Have you any letters for the place I am going to?
500	13	4	Will you take our letters for the place you are going to, or elsewhere, to be sent per address.
501	13	5	Will you take a letter or letters of importance from me for the place you are going to?
502	13	6	Will you put some letters in the post-office for me?
503	13	7	Will you call at the post-office for letters for me?
504	13	8	Will you call at the post-office for letters for the person whose name I am going to spell?
505	13	9	When did you write last? *Ans.* With date from Compound Table.
506	13	10	Do you intend to write by the first opportunity?
507	13	11	Did you write by last opportunity?
508	13	12	When do you intend to write? *Ans.* From Compound Table, date.
509	13	15	What is the date of your latest newspaper? *Ans.* From Compound Table.
510	13	16	Can you lend us your newspapers?
			ANSWERS OR COMMUNICATIONS.
511	13	17	The packet has arrived from the *place* indicated next (per Compound Table). *Note.*—At places where packets do not arrive from more places than one, no signal for place is needed, and in such case the last signal of the sentence, when answered, is hauled nearly down, then rehoisted, and struck.
512	13	18	Packet is arrived from the *place* and at the *date*, indicated by the two following pairs of signals:—(per Compound Table.) *Note.*—If there is but one packet going to the place, the signals indicating the place from which the packet comes, may be dispensed with; and in such case the last signal of the sentence, when answered, is hauled nearly down, then rehoisted, and struck; and the next pair of signals must then be understood to indicate the date.
513	13	19	Vessel in sight is a packet.
514	13	20	Vessel seen in quarter from me as indicated next (per Compound Table, point of compass) is a packet.
515	13	21	A packet is arrived from the *place* indicated next (per Compound Table).
516	13	22	The packet is now at the *place* indicated next (per Compound Table).
517	13	23	The packet brings letters down to the *date* indicated next (per Compound Table).
518	13	24	The packet is not yet arrived.

Numerical order of Sentences.	Pair Signals. No.	No.	CHAPTER VI. *Relating to Packets, Steam-boats, Mails, Letters, and Newspapers.* ANSWERS OR COMMUNICATIONS.
519	13	25	The packet was not arrived when I left.
520	13	26	The packet has sailed.
521	13	27	The packet will sail at the *date* indicated next (per Compound Table).
522	13	28	The packet has not sailed yet.
523	13	29	The packet is a steam-vessel.
524	13	30	A steam-vessel will sail at the *date,* and for the *place,* indicated by the two following pairs of signals:—(per Compound Table.)
525	13	31	The steam-boat from the *place* indicated next (per Compound Table) is not arrived.
526	13	32	A steam-boat has arrivedf rom the *place* indicated next (per Compound Table).
527	13	33	The steam-boat has not yet left.
528	13	34	There is no other regular opportunity from this.
529	13	35	There are both sailing-vessels and steam-vessels.
530	13	36	There are only sailing-vessels
531	13	37	There are only steam-vessels.
532	13	38	The mail is not arrived.
533	13	39	The mail is arrived.
534	13	40	The mail has not yet left.
535	14	1	The mail has left.
536	14	2	I have despatches for government.
537	14	3	Send for your letters.
538	14	4	I have no letters for you or any on board.
539	14	5	I have letters for you, or to your care or address.
540	14	6	There are letters for you at the post-office.
541	14	7	There are letters for the person spoken of, or whose name I shall spell. *Note.*—Where the spelling of the name may be dispensed with, the last signal of the sentence, when answered, is hauled nearly down, re-hoisted, and struck.
542	14	8	I shall take your letters to the place where I am going.
543	14	9	I have written.
544	14	10	I shall not write this time.
545	14	11	I shall write on the *date* indicated next (per Compound Table).
546	14	12	I shall write by the first opportunity.
547	14	15	I shall write by the opportunity which leaves on *date* indicated next (per Compound Table).
548	14	16	Will you write for both of us.
549	14	17	There is now an opportunity for writing home.

Numerical order of Sentences.	Pair Signals.		CHAPTER VI. *Relating to Packets, Steam-boats, Mails, Letters, and Newspapers.*
	No.	No.	ANSWERS OR COMMUNICATIONS.
550	14	18	There is now an opportunity for letters to the *place* indicated next (per Compound Table).
551	14	19	I shall deliver your letters at the post-office, or as directed.
552	14	20	I shall take up your letters and newspapers at the post-office, or as directed.
553	14	21	I have newspapers for you.
554	14	22	I have newspapers down to *date* indicated next (per Compound Table).
555	14	23	I have newspapers down to *date*, and from *place* as indicated by the two following pairs of signals:—(per Compound Table.)
556	14	24	Send for newspapers.
557.	14	25	I shall send you the letters or newspapers.

CHAPTER VII.

Numerical order of Sentences.	Pair Signals. No.	No.	*Relating to Average and Damage on Vessel and Cargo.*
			QUESTIONS.
558	14	26	Are you on shore?
559	14	27	Is vessel bearing from me as indicated next (per Compound Table, point of compass) on shore?
560	14	28	Is your vessel leaky?
561	14	29	Is the vessel we are speaking of, or which bears from me as indicated (per Compound Table, point of compass) leaky? *Note.*—If the vessel spoken of is to be understood, the last signal, when answered, is hauled nearly down, rehoisted, and struck.
562	14	30	Has your vessel been on shore?
563	14	31	Has the vessel we are speaking of, or belonging to the *place* indicated next (per Compound Table) and whose name shall be spelt, been on shore? *Note.*—See note to sentence 561.
564	14	32	Do you leak much?
565	14	33	Do you think you'll be able to keep the sea?
566	14	34	Have you received much damage?
567	14	35	Has the vessel we are speaking of, or belonging to the *place* indicated next (per Compound Table) and whose name shall be spelt, received much damage? *Note.*—See note to sentence 561.
568	14	36	What is your distress?
569	14	37	What damage has the vessel we are speaking of sustained?
570	14	38	Can vessel spoken of keep the sea?
			ANSWERS OR COMMUNICATIONS.
571	14	39	I am on shore.
572	14	40	Vessel we are speaking of, or bearing from me, as indicated next (per Compound Table, point of compass) is on shore. *Note.*—If the vessel spoken of is to be understood, the last signal, when answered, is hauled nearly down, rehoisted, and struck.
573	15	1	My vessel has not been on shore.
574	15	2	My vessel has been on shore.
575	15	3	I am afraid my vessel will not be got off.
576	15	4	Vessel we are speaking of cannot be got off.
577	15	5	My vessel is got off.
578	15	6	Vessel we are speaking of has not been on shore.

CHAPTER VII.

Relating to Average and Damage on Vessel and Cargo

Numerical order of Sentences.	Pair Signals. No.	No.	ANSWERS OR COMMUNICATIONS.
579	15	7	Vessel we are speaking of has been on shore.
580	15	8	I have sprung a leak.
581	15	9	Vessel we are speaking of has sprung a leak.
582	15	10	My vessel is not leaky.
583	15	11	Vessel we are speaking of is not leaky.
584	15	12	My vessel is very leaky, don't leave me.
585	15	13	My leak increases and is dangerous.
586	15	14	Leak is stopt.
587	15	17	My pumps are worn out and useless, can you assist me with pump and gear?
588	15	18	I have received little or no damage.
589	15	19	I have received much damage.
590	15	20	Vessel we are speaking of has received little or no damage.
591	15	21	Vessel we are speaking of has received much damage.
592	15	22	I think I can keep the sea.
593	15	23	Vessel we are speaking of thinks she can keep the sea.
594	15	24	I cannot keep the sea.
595	15	25	Vessel we are speaking of cannot keep the sea.
596	15	26	I have been obliged to throw part of my cargo overboard.
597	15	27	I have been obliged to throw my guns overboard.
598	15	28	I am floating upon my cargo of timber, come to my assistance.
599	15	29	My cargo of salt is melting away.
600	15	30	My cargo of sugar is melting away.
601	15	31	My cargo has shifted.
602	15	32	My cargo is damaged.
603	15	33	My cargo is much damaged.
604	15	34	My powder is wet.
605	15	35	I have been run on board.
606	15	36	A vessel belonging to the *place* indicated next (per Compound Table) has been run on board.
607	15	37	I have been run on board by a vessel of my own nation.
608	15	38	I have been run on board by a vessel belonging to the *place* indicated next (per Compound Table), and whose name I am going to spell. *Note.*—Should it happen that the name of the vessel is not known, then instead of spelling the name as promised in the sentence, the signal from Chap. XVII. "I don't know," is hoisted.
609	15	39	Vessel belonging to the *place* indicated next (per Compound Table), and whose name shall be spelt, has been run on board by a vessel of the *nation* indicated by the second pair of signals following:—(per Compound Table.) *Note.*—See note to the preceding sentence.

CHAPTER VII.

Relating to Average and Damage on Vessel and Cargo.

Numerical order of Sentences.	Pair Signals No.	Pair Signals No.	ANSWERS OR COMMUNICATIONS.
610	15	40	Vessel is a ship.
611	16	1	Vessel is a brig.
612	16	2	Vessel is a schooner.
613	16	3	Vessel is a galliot.
614	16	4	Vessel is a ketch.
615	16	5	Vessel is a lugger.
616	16	6	Vessel is a felucca.
617	16	7	Vessel is a cutter (yacht, sloop, or smack).
			Note.—These eight sentences may be used upon all other occasions, to denote the quality of vessel in question. For vessels of war see Chap. XI. No. 1195—1204.
618	16	8	I have received damage as indicated by signals following.
619	16	9	Vessel we are speaking of, or which belongs to the *place* indicated next (per Compound Table), and whose name shall be spelt, has received damage as indicated by signals following.
			Note.—If the vessel spoken of is to be understood, the last signal of the sentence when answered is hauled nearly down, rehoisted, and struck.—See note to 608.
620	16	10	Bowsprit is damaged.
621	16	11	Bowsprit is lost or carried away.
622	16	12	Mainmast is sprung or damaged.
623	16	13	Mainmast is lost or carried away.
624	16	14	Foremast is sprung or damaged.
625	16	17	Foremast is lost or carried away.
626	16	18	Mizen mast is sprung or damaged.
627	16	19	Mizen mast is lost or carried away.
628	16	20	Maintop mast is sprung.
629	16	21	Foretop mast is sprung.
630	16	22	Mizentop mast is sprung.
631	16	23	Topmasts are sprung.
632	16	24	Topgallant masts are sprung.
633	16	25	Mainyard is damaged.
634	16	26	Mainyard is very much damaged.
635	16	27	Foreyard is damaged.
636	16	28	Foreyard is very much damaged.
637	16	29	Maintopsail yard is sprung or carried away.
638	16	30	Foretopsail yard is sprung or carried away.
639	16	31	Damage on the smaller yards and spars.
640	16	32	Rigging is much damaged.
641	16	33	Sails are much damaged.

CHAPTER VII.

Relating to Average and Damage on Vessel and Cargo.

Numerical order of Sentences.	Pair Signals. No.	Pair Signals. No.	ANSWERS OR COMMUNICATIONS.
642	16	34	Hull is much damaged.
643	16	35	Mainsail or foresail is split or lost.
644	16	36	Courses split or lost.
645	16	37	A topsail split or lost.
646	16	38	Topsails split or lost.
647	16	39	Stormsails all split or lost.
648	16	40	Damage on or loss of smaller sails.
649	17	1	I can repair at sea.
650	17	2	I cannot repair at sea.
651	17	3	I am obliged to cut my cable, drag for it, or order it to be done, for my own, or insurer's account.
652	17	4	I am obliged to slip cables with a buoy, take it up after my departure.
653	17	5	I fear I shall part.
654	17	6	I have parted, sweep for my anchor when I am gone, expenses to my account.
655	17	7	I have parted my chain cable, sweep for it when I am gone, expenses to my account.
656	17	8	I have lost all my boats.
657	17	9	Vessel we are speaking of, or belonging to the *place* indicated next (per Compound Table), and whose name shall be spelt, has lost all her boats.
			Note.—If the vessel spoken of is to be understood, the last signal of the sentence, when answered, is hauled nearly down, rehoisted, and struck.
658	17	10	I have lost part of my crew, *number* to be indicated next (per Compound Table).
659	17	11	Vessel we are speaking of, or belonging to the *place* indicated next (per Compound Table), and whose name I shall spell, has lost part of her crew, the *number* of which shall be indicated by the second pair of signals following:—(per Compound Table.)
			Note.—If the vessel spoken of is to be understood, the last signal of the sentence, when answered is hauled nearly down, rehoisted, and struck; the pair signal expressing the number of men lost, in this case will come in immediately after the sentence. Should the number of men happen not to be exactly known, or not known at all, the signal from Chap. XVII., "I don't know," is hoisted instead of the number taken from the Compound Table.
660	17	12	I have lost all my anchors.
661	17	13	I have only one anchor left.

Numerical order of Sentences.	Pair Signals. No.	No.	CHAPTER VII. *Relating to Average and Damage on Vessel and Cargo.* ANSWERS OR COMMUNICATIONS.
662	17	14	Vessel we are speaking of, or belonging to the *place* indicated next (per Compound Table), and whose name shall be spelt, has lost all her anchors. *Note.*—If the vessel *spoken of* is to be understood, the last signal of the sentence, when answered, is hauled nearly down, rehoisted, and struck.
663	17	15	Vessel we are speaking of, or belonging to the *place* indicated next (per Compound Table), and whose name shall be spelt, has but one anchor left. *Note.*—See note to the preceding sentence.
664	17	16	Vessel in sight, or bearing from me, as indicated next (per Compound Table, point of compass), has lost all her masts. *Note.*—If the vessel in sight is to be understood, the last signal of the sentence, when answered, is hauled nearly down, rehoisted, and struck.
665	17	19	Vessel belonging to the *place*, indicated next (per Compound Table), and whose name shall be spelt, has lost all her masts.
666	17	20	Vessel in sight, or *bearing* from me, as indicated next (per Compound Table), has lost her topmasts. *Note.*—See note above to No. 664.
667	17	21	Vessel belonging to the *place* indicated next (per Compound Table), and whose name shall be spelt, has lost her topmasts.
668	17	22	Vessel in sight, or *bearing* from me, as indicated next (per Compound Table), is quite a wreck. *Note.*—See the preceding notes.
669	17	23	Vessel belonging to the *place* indicated next (per Compound Table), and whose name shall be spelt, has got into the *port* indicated by the second pair of signals following (per Compound Table), in a sinking state, or as a wreck.
670	17	24	Vessel belonging to the *place* indicated next (per Compound Table), and whose name shall be spelt, is lost or foundered at sea, or at the *place* indicated by the second pair of signals following:—(per Compound Table.) *Note.*—If lost at sea, the name of the sea will be found in the Compound Table among the places, &c.
671	17	25	Vessel belonging to the *place* indicated next (per Compound Table), and whose name shall be spelt, has put into the *port* indicated by the second pair of signals following (per Compound Table), under average.

Numerical order of Sentences.	Pair Signals. No.	Pair Signals. No.	CHAPTER VII. *Relating to Average and Damage on Vessel and Cargo.* ANSWERS OR COMMUNICATIONS.
672	17	26	All on board saved.
673	17	27	Captain alone saved
674	17	28	All saved except the captain.
675	17	29	Part of the crew saved.
676	17	30	Nobody saved.
677	17	31	It is not known if crew were saved or not.
678	17	32	Vessel we are speaking of, or belonging to the *place* indicated next (per Compound Table), and whose name shall be spelt, has never been heard of. *Note.*—If the vessel spoken of is to be understood, the last signal of the sentence, when answered, is hauled nearly down, rehoisted, and struck.
679	17	33	The *number* of vessels indicated next (per Compound Table) were lost in the gale or hurricane we are speaking of, or in the last gale or hurricane. *Note.*—See note above.
680	17	34	I have discovered, or I see people on a wreck in the *point of compass* from me as indicated next (per Compound Table).
681	17	35	Rudder is damaged.
682	17	36	Rudder is lost.
683	17	37	Vessel belonging to the *place* indicated next (per Compound Table), and whose name shall be spelt, has lost her rudder.

CHAPTER VIII.

Numerical order of Sentences.	Pair Signals. No.	No.	*Relating to Assistance in general, sparing and lending different things.* QUESTIONS.
684	17	38	I have lost part of my crew, can you assist me with some hands, and how many? *Ans.* From Compound Table, or from Chap. XVII., "I cannot comply with your request, &c."
685	17	39	Can you to-day assist me with hands, *number* to be indicated next (per Compound Table)?
686	17	40	Can you lend me your boatswain?
687	18	1	Can you lend me your gunner?
688	18	2	Can you lend me your carpenter?
689	18	3	Can you lend me your sailmaker?
690	18	4	Can you lend me your cooper?
691	18	5	Can you lend me some hands who can paint?
692	18	6	I cannot purchase my anchor without help; send me hands.
693	18	7	Will you take me in tow?
694	18	8	Shall I take you in tow?
695	18	9	Will you take vessel, *bearing* from me as indicated (per Compound Table), in tow?
696	18	10	Can you assist me?
697	18	11	Do you want assistance? *Ans.* No; or according to circumstances.
698	18	12	Can you give assistance to vessel *bearing* from me, as indicated next (per Compound Table)?
699	18	13	Does vessel *bearing* from me, as indicated next (per Compound Table), want assistance?
700	18	14	What assistance do you want?
701	18	15	What assistance does vessel *bearing* from me, as indicated next (per Compound Table), want?
702	18	16	Shall I order you assistance from shore?
703	18	19	Shall I report your damage, and where must I report it? *Ans.* No; or place from Compound Table.
704	18	20	Will you report the damage I have sustained?
705	18	21	Will you send me an anchor and cable?
706	18	22	Send an anchor and cable to the vessel *bearing* from me, as indicated next (per Compound Table, point of compass).
707	18	23	Can you lend me your long boat?
708	18	24	Can you lend me a boat?
709	18	25	Make sail a-head, and drop a boat on board.

Numerical order of Sentences.	Pair Signals. No.	No.	CHAPTER VIII. *Relating to Assistance in general, sparing and lending different things.* QUESTIONS.
710	18	26	Send a boat to my assistance.
711	18	27	Send boats to assistance of vessel *bearing* from me, as indicated next (per Compound Table).
712	18	28	Send boats with hawsers to my assistance.
713	18	29	Send boats with hawsers to vessel *bearing* from me, as indicated next (per Compound Table).
714	18	30	Send my boat a warp, as it is difficult for it to pull up.
715	18	31	Have you a chart of the *sea, or of the entrance to place* indicated next (per Compound Table)?
716	18	32	Can you lend me a chart of the *sea, or of the entrance* to place indicated next (per Compound Table)?
717	18	33	I am lying becalmed, and current setting on shore; send boats to tow.
718	18	34	Vessel *bearing* from me, as indicated next (per Compound Table), is becalmed, and current setting on shore; send boats to tow.
719	18	35	I require immediate assistance.
720	18	36	Vessel we are speaking of, or *bearing* from me, as indicated next (per Compound Table), wants immediate assistance.
			Note.—If vessel spoken of is understood, the last signal, when answered, is hauled nearly down, rehoisted, and struck.
721	18	37	Will you send for it, or for them?
722	18	38	Will you send for the person?
723	18	39	Can you sell or lend me article indicated next?
724	18	40	Can you sell or lend me the articles indicated next?
			Note.—If nothing can be spared, the answer may be simply, No; but if the request can be complied with, one of the following signals is made according to circumstances.
725	19	1	How many bottles do you want?
726	19	2	How many gallons do you want?
727	19	3	How many pounds do you want?
728	19	4	How many pieces do you want?
729	19	5	How many barrels do you want?
730	19	6	How many feet do you want?
731	19	7	Of what length in feet and diameter in inches must it be?
			Note.—The above seven sentences are answered by indicating the quantity simply by numbers (per Compound Table), and the last of them requires two pairs of signals, one for the length and the other for the diameter.
			At the end of this chapter will be found a list of such things and articles as a vessel is most likely to be in want of; should a thing wanted not be found in this list recourse must be had to spelling

Numerical order of Sentences.	Pair Signals. No.	No.	CHAPTER VIII. *Relating to Assistance in general, sparing and lending different things.* ANSWERS OR COMMUNICATIONS.
732	19	8	I cannot spare more hands than the *number* indicated next (per Compound Table).
733	19	9	I cannot assist you with men, or the number of men you request.
734	19	10	Send for the men you want.
735	19	11	I shall send you assistance of men.
736	19	12	I shall assist you as far as I can.
737	19	13	I cannot assist you.
738	19	14	I shall take you in tow.
739	19	15	I shall take vessel in question in tow.
740	19	16	I cannot take your vessel in tow.
741	19	17	I cannot take vessel indicated in tow.
742	19	18	I shall send you, or vessel in question, anchor and cable.
7 3	19	21	Send for anchor and cable.
744	19	22	I have no long boat.
745	19	23	My boat is stove.
746	19	24	All my boats are lost or stove.
747	19	25	I shall send you a boat.
748	19	26	I have no boat on board.
749	19	27	I shall send hawsers to your assistance, or assistance of vessel in question.
750	19	28	Send for hawsers to you or vessel in question.
751	19	29	I have no chart of the sea named to spare.
752	19	30	I shall send you, or vessel in question, boats to tow.
753	19	31	You may send for person, thing, or article, indicated next.
754	19	32	Send on board for person, thing, or article, indicated next.
755	19	33	Send on shore for the thing wanted, or person, thing, or article indicated next. *Note.*—If the thing wanted is to be understood, the last signal, when answered, is hauled nearly down, rehoisted, and struck.
756	19	34	Send immediately.
757	19	35	I shall send for it.
758	19	36	I shall send for the person.
759	19	37	Send for the thing wanted.
760	19	38	I shall send you the thing wanted
761	19	39	I shall send for the thing I want.
762	19	40	I cannot send.

CHAPTER VIII.

List of Ship Furniture, Stores, Provisions, Cabin Inventory, Ammunition and Arms, Books and Instruments, Medicines, &c.

Numerical order of Sentences.	Pair Signals. No.	No.	SHIP FURNITURE.
763	20	1	Anchor (heavy)
764	20	2	Anchor and cable.
765	20	3	Kedge anchor.
766	20	4	Anchor stock.
767	20	5	Small anchor (grapnel)
768	20	6	Long boat.
769	20	7	Boat (in general).
770	20	8	Boat oars.
771	20	9	Buoy.
772	20	10	Hawsers.
773	20	11	Hawsers, small.
774	20	12	Capstan bars.
775	20	13	Buckets.
776	20	14	Scrubbing-brushes, deck scrubbers.
777	20	15	Brooms.
778	20	16	Tar brushes.
779	20	17	Paint brushes.
780	20	18	Tar blacking, black paint.
781	20	21	New canvass.
782	20	22	Old canvass.
783	20	23	Twine.
784	20	24	Bunting.
785	20	25	Colour or national flag of my own nation.
786	20	26	Colour or national flag of *nation* indicated next (per Compound Table).
787	20	27	Jack of my own nation.
788	20	28	Jack of the *nation* indicated next (per Compound Table).
789	20	29	Sail-needle and palm.
790	20	30	Handspikes.
791	20	31	Iron crowbar.
792	20	32	Treble blocks.
793	20	33	Double blocks.
794	20	34	Single blocks.
795	20	35	Snatch block.
796	20	36	Dead eyes.
797	20	37	Thimbles.
798	20	38	Tackles: double and single block
799	20	39	Tackles: two double blocks.
800	20	40	Tackles: treble and double, or two treble blocks, winding tackle.
801	21	1	Bull's eyes (patent illuminators).
802	21	2	Glass panes.

CHAPTER VIII.

List of Ship Furniture, Stores, Provisions, Cabin Inventory, Ammunition and Arms, Books and Instruments, Medicines, &c.

Numerical order of Sentences.	Pair Signals No.	Pair Signals No.	STORES.
803	21	3	Hammocks for men.
804	21	4	Hammocks for passengers (cots).
805	21	5	Hose.
806	21	6	Glue.
807	21	7	Grindstone.
808	21	8	Files.
809	21	9	Hammer.
810	21	10	Plane.
811	21	11	Axe.
812	21	12	Adze.
813	21	13	Chisels.
814	21	14	Saw.
815	21	15	Chalk.
816	21	16	Nails, large.
817	21	17	Nails, small.
818	21	18	Nails, sheathing.
819	21	19	Pump tacks.
820	21	20	Pump plat heads, or scupper nails.
821	21	23	Copper sheet.
822	21	24	Carpenters' tools in general.
823	21	25	Raw hides.
824	21	26	Iron, bars for manufacturing.
825	21	27	Coopers' adze.
826	21	28	Staves.
827	21	29	Iron hoops.
828	21	30	Wooden hoops.
829	21	31	Coopers' tools in general.
830	21	32	Rope, new.
831	21	33	Rope, old.
832	21	34	Lines, tarred, or not tarred.
833	21	35	Oakum.
834	21	36	Marline spikes.
835	21	37	Spun yarn.
836	21	38	Grease.
837	21	39	Pitch.
838	21	40	Tar.
839	22	1	Rosin.
840	22	2	Turpentine.
841	22	3	Lamp.
842	22	4	Lamp oil.
843	22	5	Paint oil.
844	22	6	Paint, black.

CHAPTER VIII.

List of Ship Furniture, Stores, Provisions, Cabin Inventory, Ammunition and Arms, Books and Instruments, Medicines, &c.

Numerical order of Sentences.	Pair Signals. No.	No	STORES.
845	22	7	Paint, blue.
846	22	8	Paint, green.
847	22	9	Paint, red.
848	22	10	Paint, white.
849	22	11	Paint, yellow.
850	22	12	Tarpauling.
851	22	13	Signal lantern.
852	22	14	Lead, sheet.
853	22	15	Lead, pig.
854	22	16	Line, deep sea.
855	22	17	Line, log.
856	22	18	Line, fishing, and hooks.
857	22	19	Fishing-nets.
858	22	20	Harpoon.
859	22	23	Scrapers.
860	22	24	Lime.
861	22	25	Mahogany wood, log or plank.
862	22	26	Oak plank.
863	22	27	Spars, fir or pine.
864	22	28	Fir plank.
865	22	29	Fir deals.
866	22	30	Shovels.
867	22	31	Spades.
868	22	32	Barrel slings.
869	22	33	Can hooks.
870	22	34	Sand.
871	22	35	Topmast, main or fore.
872	22	36	Ditto mizen.
873	22	37	Mainyard, or fore ditto.
874	22	38	Topsail yard.
875	22	39	Spar for lesser mast or yard.
876	22	40	Mainsail or fore ditto.
877	23	1	Topsail, main or fore.
878	23	2	Ditto mizen.
879	23	3	Topgallant sail.
880	23	4	Mizen (spanker driver).
881	23	5	Jib.
882	23	6	Fore staysail.
883	23	7	Lower studding-sail.
884	23	8	Topmast ditto.

Numerical order of Sentences.	Pair Signals. No.	No.	CHAPTER VIII. *List of Ship Furniture, Stores, Provisions, Cabin Inventory, Ammunition and Arms, Books and Instruments, Medicines, &c.* PROVISIONS.
885	23	9	Beef, fresh.
886	23	10	Beef, salt.
887	23	11	Pork, salt.
888	23	12	Pork, fresh, or pigs.
889	23	13	Fish, fresh (for salt fish, and stock-fish, see Chap. XIII.)
890	23	14	Herrings, salted.
891	23	15	Bread (biscuits).
892	23	16	Soft bread.
893	23	17	Rice.
894	23	18	Barley grits.
895	23	19	Pease.
896	23	20	White beans.
897	23	21	Spirits (brandy, gin, whiskey).
898	23	22	Rum.
899	23	25	Butter.
900	23	26	Hogs' lard.
901	23	27	Cheese.
902	23	28	Cocoa (beans).
903	23	29	Chocolate.
904	23	30	India corn.
905	23	31	India meal.
906	23	32	Tea.
907	23	33	Coffee.
908	23	34	Sugar, brown.
909	23	35	Sugar, lump.
910	23	36	Eggs.
911	23	37	Flour.
912	23	38	Rye meal.
913	23	39	Milk.
914	23	40	Sheep or lamb.
915	24	1	Mustard.
916	24	2	Oats.
917	24	3	Hay.
918	24	4	Sweet oil.
919	24	5	Onions.
920	24	6	Oranges.
921	24	7	Lemon, limes, or juice
922	24	8	Black pepper.
923	24	9	Cayenne pepper.
924	24	10	Pickles.
925	24	11	Porter.
926	24	12	Beer.

Numerical order of Sentences.	Pair Signals. No.	No.	CHAPTER VIII. *List of Ship Furniture, Stores, Provisions, Cabin Inventory, Ammunition and Arms, Books and Instruments, Medicines, &c.*
			PROVISIONS.
927	24	13	Potatoes.
928	24	14	Poultry.
929	24	15	Dried raisons.
930	24	16	Sweet almonds.
931	24	17	Table salt.
932	24	18	Vinegar.
933	24	19	Spices.
934	24	20	Greens.
935	24	21	Water.
936	24	22	Madeira wine.
937	24	25	Port ditto.
938	24	26	Malaga ditto, or sweet wines in general.
939	24	27	Bordeaux ditto (claret).
940	24	28	Sherry (or Teneriffe).
941	24	29	Champagne.
			MEDICINES.
942	24	30	China bark (Peruvian bark, quinquina), quinine.
943	24	31	Calomel.
944	24	32	Lint.
945	24	33	Salts (purgative).
946	24	34	Lancet.
947	24	35	Blister.
948	24	36	Clyster-pipe (syringe).
949	24	37	Castor oil (purgative).
950	24	38	Oat grits.
951	24	39	Sago, or pearl barley.
952	24	40	Saloop (Indian arrow-root)
953	25	1	Sour krout.
954	25	2	Medicines in general.
			CABIN INVENTORY.
955	25	3	China, or earthenware dishes.
956	25	4	Tureen.
957	25	5	Plates.
958	25	6	Teapot.
959	25	7	Teacups and saucers. or basins.
960	25	8	Table spoons.
961	25	9	Tea spoons.
962	25	10	Knives and forks.
963	25	11	Wine or beer glasses.

Numerical order of Sentences.	Pair Signals. No.	No.	CHAPTER VIII. *List of Ship Furniture, Stores, Provisions, Cabin Inventory, Ammunition and Arms, Books and Instruments, Medicines, &c.*
			CABIN INVENTORY, Continued.
964	25	12	Tea-kettle.
965	25	13	Cruet-stand (plat de ménage).
			ARMS AND AMMUNITION.
966	25	14	Matches.
967	25	15	Cannon-balls (weight indicated next, per Compound Table, number of pounds).
968	25	16	Great guns.
969	25	17	Muskets with ammunition.
970	25	18	Pistols with ammunition.
971	25	19	Cutlasses.
972	25	20	Boarding-pikes.
973	25	21	Rockets.
974	25	22	Blue lights.
975	25	23	Gunpowder.
976	25	24	Gunpowder for small arms.
977	25	27	Grape-shot, or canister (weight indicated next, per Compound Table, number of pounds).
978	25	28	Musket flints.
979	25	29	Ammunition in general.
			BOOKS, OR NAUTICAL INSTRUMENTS.
980	25	30	Nautical almanack for year indicated next, (number from Compound Table).
			Note.—As the numbers greater than 1547 would require two pairs of signals; to save time the year 1835, for example, may be given simply 35, from Compound Table.
981	25	31	Log-book, blank.
982	25	32	Books, reading, for entertainment.
983	25	33	Chronometer.
984	25	34	Barometer.
985	25	35	Thermometer.
986	25	36	Steering compass.
987	25	37	Azimuth compass (or for bearings).
988	25	38	Sextant.
989	25	39	Quadrant.
990	25	40	Spy-glass.
991	26	1	Pair of compasses, case of mathematical drawing instruments.

CHAPTER VIII.

List of Ship Furniture, Stores, Provisions, Cabin Inventory, Ammunition and Arms, Books and Instruments, Medicines, &c.

Numerical order of Sentences.	Pair Signals. No.	Pair Signals. No.	SUNDRIES.
992	26	2	Canvass sacks.
993	26	3	Scales and Weights.
994	26	4	Steelyard.
995	26	5	Water casks.
996	26	6	Water barrels.
997	26	7	Hand pump.
998	26	8	Kegs, ankers, (empty).
999	26	9	Leather for shoes.
1000	26	10	Woollen blankets.
1001	26	11	Clothing for the crew.
1002	26	12	Cotton.
1003	26	13	Bottles, empty.
1004	26	14	Demi-johns.
1005	26	15	Bottle corks.
1006	26	16	Sea coals.
1007	26	17	Charcoal.
1008	26	18	Wood for fuel.
1009	26	19	Gallon measure.
1010	26	20	Mattresses.
1011	26	21	Needles and thread.
1012	26	22	Soap.
1013	26	23	Razor and shaving things.
1014	26	24	Water funnel.
1015	26	27	Writing-paper.
1016	26	28	Writing pens, quills.
1017	26	29	Ink.
1018	26	30	Tallow candles.
1019	26	31	Wax or spermaceti candles
1020	26	32	Tobacco-pipes.
1021	26	33	Tobacco for smoking.
1022	26	34	Cigars.
1023	26	35	Tobacco for chewing.

CHAPTER IX.

Numerical order of Sentences.	Pair Signals. No.	No.	*Relating to Commercial Affairs.* QUESTIONS.
1024	26	36	Who is your agent, or to whom are you consigned? *Ans.* By spelling the name or firm of the house. *Note.*—See directions for spelling and abbreviating.
1025	26	37	Who is the agent for Lloyd's at this place? *Ans.* As above.
1026	26	38	Will you report to my owner my vessel sails well?
1027	26	39	Will you report to my owner, or persons concerned, that I have met a good market?
1028	26	40	Report my vessel where you come, all well on board.
1029	27	1	Is there freight to be had at the place you come from? *Ans.* No; or according to circumstances.
1030	27	2	Is there freight to be had at the *place* indicated next (per Compound Table)? *Ans.* No; or according to circumstances.
1031	27	3	At what place do you think it most likely for me to procure a freight? *Ans.* From Compound Tab.e, name of place, or Chap. XVII.
1032	27	4	I wish to sell my vessel; what place would you recommend me to go to, as most likely to meet with an advantageous sale? *Ans.* From Compound Table, name of place, or Chap. XVII.
1033	27	5	How many dollars and cents go to a pound sterling by the last course of exchange? *Ans.* From Compound Table, with number for cents; as, for example, the pound sterling being equal to 4 dollars 80 cents (which is also written $4.80), the signal for 480 is made.
1034	27	6	How many shillings banco go to a pound sterling by the last course of exchange? *Ans.* From Compound Table, number.
1035	27	7	How many Dutch guilders go to a pound sterling by the last course of exchange? *Ans.* From Compound Table by a pair signal for the number of guilders, and another expressing the number of stivers.

Numerical order of Sentences	Pair Signals. No.	No.	CHAPTER IX. *Relating to Commercial Affairs.* QUESTIONS.
1036	27	8	How many francs go to a pound sterling by the last course of exchange? *Ans.* As the preceding sentence, with a pair signal for the francs, and another for the centimes, per Compound Table.
1037	27	9	How many per cents are paid for silver dollars against Spanish doubloons in gold? *Ans.* From Compound Table.
1038	27	10	How many Rigsbank shillings (Danish currency) go to the species dollar by the last course of exchange? *Ans.* From Compound Table.
1039	27	11	At the last exchange what was the value of a pound sterling in the currency I am going to indicate by spelling? *Ans.* From Compound Table.
1040	27	12	What is the price of coffee per 100 lb. English, at the place you are coming from? *Ans.* From Compound Table. *Note.*—The price is given in cents of dollars, in answer to this and the following sentences of the same nature; for example, the price of coffee is to be signalized at 11 dollars and 40 cents—then 1140 is signalized from Compound Table, numbers: when the number exceeds 1547, or 15 dollars and 47 cents, see Introduction to the Compound Table for directions in such cases.
1041	27	13	What is the price of white Havannah sugar per 100 lb. English, at the place you are coming from? *Ans.* As above. Number of cents of Spanish dollars.
1042	27	14	What is the price of brown Havannah sugar per 100 lb. English, at the place you are coming from? *Ans.* As above. Number of cents of Spanish dollars.
1043	27	15	What is the price of raw sugars per 100 lb. English, at the place you are coming from? *Ans.* As above. Number of cents of Spanish dollars.
1044	27	16	What is the price of rum per gallon, at the place you are coming from? *Ans.* From Compound Table. Number of cents of Spanish dollars.
1045	27	17	What is the price of cotton per pound, English, at the place you are coming from? *Ans.* From Compound Table as above. Number of cents of Spanish dollars.

Numerical order of Sentences.	Pair Signals.		CHAPTER IX. *Relating to Commercial Affairs.*
	No.	No.	QUESTIONS.
1046	27	18	What is the price of flour per barrel, at the place you are coming from? *Ans.* From Compound Table as above. Number of cents of Spanish dollars.
1047	27	19	What is the price of rye-meal per barrel, at the place you are coming from? *Ans.* From Compound Table as above. Number of cents of Spanish dollars.
1048	27	20	What is the price of India meal per barrel, at the place you are coming from? *Ans.* From Compound Table as above. Number of cents of Spanish dollars.
1049	27	21	What is the price of India corn per barrel, at the place you are coming from? *Ans.* From Compound Table as above. Number of cents of Spanish dollars.
1050	27	22	What is the price of Havannah cigars per thousand, at the place you are coming from? *Ans.* From Compound Table as above. Number of cents of Spanish dollars.
1051	27	23	What is the price of indigo per pound, English, at the place you are coming from? *Ans.* From Compound Table as above. Number of cents of Spanish dollars.
1052	27	24	What is the price of salted herrings per barrel, at the place you are coming from? *Ans.* From Compound Table as above. Number of cents of Spanish dollars.
1053	27	25	What is the price of fustic per ton, English, at the place you are coming from? *Ans.* From Compound Table as above. Number of cents of Spanish dollars.
1054	27	26	What is the price of logwood per ton, English, at the place you come from? *Ans.* From Compound Table as above. Number of cents of Spanish dollars.
1055	27	29	What is the price of tobacco per 100 lb. English, at the place you are coming from? *Ans.* From Compound Table as above. Number of cents of Spanish dollars.
1056	27	30	What is the price of rice per 100 lb. English, at the place you are coming from? *Ans.* From Compound Table as above. Number of cents of Spanish dollars.
1057	27	31	What is the price of coals per keel, English, at the place you come from? *Ans.* From Compound Table, by a pair of signals for dollars, and another for cents.

Numerical order of Sentences.	Pair Signals. No.	No.	CHAPTER IX. *Relating to Commercial Affairs.*
			QUESTIONS.
1058	27	32	What is the price of salt per ton, English, at the place you come from? *Ans.* From Compound Table, by a pair of signals, for dollars, and another for cents.
1059	27	33	What is the price of tar per barrel at the place you are coming from? *Ans.* From Compound Table, by number of cents of Spanish dollars.
1060	27	34	What is the price of flax per 100 lb., English, at the place you are coming from? *Ans.* From Compound Table by number of dollars, and another pair of signals for cents.
1061	27	35	What is the price of hemp per ton, English, at the place you are coming from? *Ans.* From Compound Table as above. Number of dollars, and next the number of cents.
1062	27	36	What is the price of iron per ton, English, at the place you are coming from? *Ans.* From Compound Table, as above, number of dollars, and next the number of cents.
1063	27	37	How does the crop or harvest promise at the place you are coming from?
1064	27	38	How does the crop or harvest promise at the *place* indicated next (per Compound Table)?
1065	27	39	How is the market at the place you are coming from, or *place* indicated next (per Compound Table)? *Note.*—If the place the other is coming from is to be understood, the last signal, when answered, is hauled nearly down, rehoisted, and struck.
1066	27	40	Does the cargo signalized find a good market at the place you are coming from, or *place* indicated next (per Compound Table)? *Note.*—By cargo signalized is here understood cargo or article spoken of, or already named in the foregoing conversation. See also note to the preceding sentence.
			ANSWERS OR COMMUNICATIONS.
1067	28	1	Transports are wanted at the place I am coming from.
1068	28	2	Good freights are to be had at the place I am coming from, or place spoken of.
1069	28	3	I am not consigned to any body.
1070	28	4	A bad crop or harvest is expected.
1071	28	5	A middling crop or harvest is expected.
1072	28	6	A good crop or harvest is expected.

Numerical order of Sentences.	Pair Signals.		CHAPTER IX. *Relating to Commercial Affairs.*
	No.	No.	ANSWERS OR COMMUNICATIONS.
1073	28	7	Trade is bad at the place I am coming from, or place in question.
1074	28	8	Market is good at the place I am coming from, or place in question.
1075	28	9	The house or firm which I shall spell, has failed.
1076	28	10	The house or firm at the *place* indicated next (per Compound Table), and whose name I shall spell, has failed.

CHAPTER X

Numerical order of Sentences.	Pair Signals. No.	No.	*Relating to Political News.*
			QUESTIONS.
1077	28	11	Has blockade been taken off at *place* indicated next (per Compound Table)?
1078	28	12	Will blockade soon be taken off at the place you come from, or *to be* indicated next (per Compound Table)? *Note.*—If the place vessel comes from is to be understood, the last signal, when answered, is hauled nearly down, rehoisted, and struck.
1079	28	13	What news about the war?
1080	28	14	What news from *country* indicated next (per Compound Table)?
1081	28	15	Has the fleet or squadron at *place or country* indicated next (per Compound Table) put to sea?
1082	28	16	What news about the fleet or squadron in *sea,* indicated next (per Compound Table)?
			ANSWERS OR COMMUNICATIONS.
1083	28	17	The place I am coming from, or *place* indicated next (per Compound Table), is blockaded by the *nation* indicated by the second pair of signals following:—(per Compound Table.) *Note.*—If place vessel is coming from is understood, the last signal of the sentence, when answered, is hauled nearly down, rehoisted, and struck.
1084	28	18	Blockade is taken off the place I am coming from, or *which* will be indicated next (per Compound Table). *Note.*—See preceding note.
1085	28	19	Blockade will soon be taken off the place I am coming from, or *which* will be indicated next (per Compound Table). *Note.*—See notes preceding.
1086	28	20	Great warlike preparations are made at the *place* indicated next (per Compound Table).
1087	28	21	The admiral whose name shall be spelt, is recalled from his command.
1088	28	22	The general whose name shall be spelt, has been recalled from his command.

Numerical order of Sentences.	Pair Signals. No.	No.	CHAPTER X. *Relating to Political News.* ANSWERS OR COMMUNICATIONS.
1089	28	23	The fleet or squadron at the *place* indicated next (per Compound Table), is ice-bound.
1090	28	24	The fleet or squadron at the *place* indicated next (per Compound Table) has put to sea—further destination unknown.
1091	28	25	The fleet or squadron at the *place* indicated next (per Compound Table) has put to sea; it is said to be destined for *place* indicated by the second pair of signals following:—(per Compound Table.)
1092	28	26	The fleet or squadron at the *place* indicated next (per Compound Table) has put to sea, and it is said is going to join the fleet or squadron of the *nation*, indicated by the second pair of signals following:—(per Compound Table.)
1093	28	29	Embargo is laid on vessels of your nation at the place I am coming from, or at *place* indicated by next pair of signals (per Compound Table). *Note.*—If place vessel is coming from is to be understood, the last signal of the sentence, when answered, is hauled nearly down, rehoisted, and struck.
1094	28	30	Embargo is laid on vessels of the *nation* indicated next (per Compound Table), at the place I am coming from, or *place* indicated by my second pair of signals following:—(per Compound Table.) *Note.*—See preceding note.
1095	28	31	Embargo has been taken off at the place I am coming from, or *place* indicated next (per Compound Table). *Note.*—See notes preceding.
1096	28	32	Emperor of —— (to be indicated by next pair of signals, per Compound Table, for nation).
1097	28	33	Empress of —— ditto.
1098	28	34	King of —— ditto.
1099	28	35	Queen of —— ditto.
1100	28	36	President of —— ditto.
1101	28	37	Pope.
1102	28	38	Crown Prince of —— ditto.
1103	28	39	Crown Princess of —— ditto.
1104	28	40	Prince of —— ditto.
1105	29	1	Princess of —— ditto.
1106	29	2	Duke of —— ditto.
1107	29	3	Duchess of —— ditto.
1108	29	4	Prime Minister of —— ditto.
1109	29	5	Minister of War of —— ditto

Numerical order of Sentences.	Pair Signals. No.	No.	CHAPTER X. *Relating to Political News.* ANSWERS OR COMMUNICATIONS.
1110	29	6	Minister of Marine or Naval Department of — (to be indicated by next pair of signals, per Compound Table, for nation).
1111	29	7	Minister of Finances of —— ditto.
1112	29	8	Minister of Foreign Affairs of —— ditto.
1113	29	9	The whole Ministry of —— ditto.
1114	29	10	The minister whose name shall be spelt next *Note.*—This sentence may be made use of to indicate any minister of a department not mentioned in the preceding sentences.
1115	29	11	is (or about to be) married to ——.
1116	29	12	is, or are dead.
1117	29	13	is, or are to be replaced (changed).
1118	29	14	Internal disturbances have broken out in the *country* indicated next (per Compound Table).
1119	29	15	Parliament (congress, chambers, diet, &c., according to the different constitutional appellation) of the *country* indicated next (per Compound Table), has been convoked to *date* indicated by the second pair of signals following:—(per Compound Table.)
1120	29	16	Parliament (congress, chamber, &c., see above), in the *country* indicated next (per Compound Table), has been dissolved.

CHAPTER XI.

Numerical order of Sentences.	Pair Signals. No.	No.	*Relating to War, Fighting, &c.* QUESTIONS.
1121	29	17	Shall I commence action?
1122	29	18	Can you renew action?
			Ans. By Yes, or No.
1123	29	19	How many killed and wounded have you?
			Ans. From Compound Table in the same order.
			Note.—If none are killed, the signal indicating none, or O, is hoisted first, and next follows the pair signal for the number of wounded, or according to circumstances. But, if neither killed nor wounded, the answer may be taken from Chap. XVII: "I have none, or, I have had none."
1124	29	20	Have you any ammunition left?
			Ans. By Yes, or No.
1125	29	21	Have you been attacked?
1126	29	22	Has vessel belonging to *place* indicated next (per Compound Table), and whose name shall be spelt, been attacked?
1127	29	23	Shall we hoist colours, and what colours shall we show?
			Ans. By No, or by signal from Compound Table, indicating the nation; or an answer may be found under the following answering sentences of this chapter.
1128	29	24	What colours does vessel in question show, or vessel *bearing* from me as indicated next (per Compound Table).
			Note.—If vessel in question is to be understood, the last signal of the sentence, when answered, is hauled nearly down, rehoisted, and struck.
1129	29	25	Have you met, or seen, any fleet, squadron, convoy, or transports at sea, and how did it, or they steer?
1130	29	26	Have you made any captures?
1131	29	27	Has vessel belonging to *place* indicated next (per Compound Table), and whose name shall be spelt, made any captures?
1132	29	28	Shall we give chase to vessel in sight, or bearing from me, as indicated next (per Compound Table, point of compass)?
			Ans. By Yes, or No.
			Note.—If vessel in sight is to be understood, the last signal, when answered, is hauled nearly down, rehoisted, and struck.

Numerical order of Sentences.	Pair Signals. No.	No.	CHAPTER XI. *Relating to War, Fighting, &c.* QUESTIONS.
1133	29	31	Are you armed ? *Ans.* By Yes, or No.
1134	29	32	Is vessel spoken of, or vessel belonging to *place* indicated next (per Compound Table), and whose name shall be spelt, armed ? *Ans.* By Yes or No, or by a sentence from this chapter. *Note.*—If vessel spoken of is to be understood, the last signal of the sentence, when answered, is hauled nearly down, rehoisted, and struck.
1135	29	33	Have you long guns, or carronades ?
1136	29	34	How many short guns or carronades, and how many long guns have you ? *Ans.* By two pair signals, for the number of each.
1137	29	35	What calibre have your guns ? *Ans.* By number from Compound Table indicating the weight of the ball.
1138	29	36	How is vessel spoken of, or vessel belonging to *place* indicated next (per Compound Table.), and whose name shall be spelt, armed ? *Ans.* By number of guns, from Compound Table, or "I don't know," &c., from Chap. XVII. *Note.*—See note to sentence 1134.
1139	29	37	Do you know what calibre vessel spoken of, or vessel belonging to *place* indicated next (per Compound Table), and whose name shall be spelt, mounts ? *Ans.* By number from Compound Table, indicating weight of the ball, or by "I don't know," &c. from Chap. XVII. *Note.*—See note to sentence 1134.
1140	29	38	Do you take vessel bearing, as indicated next (per Compound Table, point of compass), to be a man-of-war ?
1141	29	39	What rate is she ? *Ans.* From the answering sentences of this chapter, whether line of battle ship, frigate, &c.
1142	29	40	Who commands the man-of-war, whose name shall be spelt next ? *Ans.* By spelling the name of the Captain, or by "I don't know," from Chap. XVII.
1143	30	1	Who is commanding officer on this station ?

CHAPTER XI.

Relating to War, Fighting, &c.

Numerical order of Sentences.	Pair Signals No.	Pair Signals No.	
			QUESTIONS.
1144	30	2	Who is commanding officer on *station* indicated next (per Compound Table)? *Note.*—This and the preceding sentence is answered, by spelling the name of the officer, or by, "I don't know," from Chap. XVII.
1145	30	3	By whom is privateer, belonging to *place* indicated next (per Compound Table), and whose name shall be spelt, commanded? *Ans.* By spelling the name, or by "I don't know," from Chap. XVII.
			ANSWERS OR COMMUNICATIONS.
1146	30	4	I shall commence action.
1147	30	5	Commence action.
1148	30	6	I have but little ammunition left, can you spare me some?
1149	30	7	I have been attacked by an enemy's privateer.
1150	30	8	Vessel belonging to *place* indicated next (per Compound Table), and whose name shall be spelt, has been attacked by enemy's privateer.
1151	30	9	I have had a running fight with a privateer.
1152	30	10	I have had a running fight with a man-of-war.
1153	30	11	Vessel belonging to *place* indicated next (per Compound Table), and whose name shall be spelt, has had a running fight with a privateer, or armed vessel.
1154	30	12	I have no other colours but those of my own nation.
1155	30	13	Let us show colours.
1156	30	14	Vessel has shown colours.
1157	30	15	Vessel has shown colours of the *nation* indicated next (per Compound Table).
1158	30	16	Vessel has struck her colours.
1159	30	17	I have met a fleet or squadron, of *so many* sail as will be indicated by my first pair of signals following, and their *course* will next be indicated (both by Compound Table).
1160	30	18	I have met a convoy of *so many* sail, as will be indicated by my first pair of signals following, and the *course* they were steering will next be indicated (both from Compound Table).
1161	30	19	In the convoy signalized were *as many* men of war as will be indicated next (per Compound Table).
1162	30	20	The fleet, squadron, convoy, or transports were met with by me near the *place* indicated next (per Compound Table),

Numerical order of Sentences.	Pair Signals. No.	No.	CHAPTER XI. *Relating to War, Fighting, &c.* ANSWERS OR COMMUNICATIONS.
			and on the *date* which the second pair of signals following will indicate (per Compound Table).
1163	30	21	The fleet, squadron, convoy, or transport, were met with by me in *latitude* and *longitude* as indicated next (per Compound Table).
1164	30	22	The ships of war were of the *nation* indicated next (per Compound Table).
1165	30	23	The ships of war were of the *nations* indicated by the following pairs of signals:—(per Compound Table.)
1166	30	24	I cannot tell of what nation the men-of-war were, as they showed no colours, neither could I make it out from their appearance.
1167	30	25	Transports are said to have on board the *number* of troops, as indicated next (per Compound Table.)
1168	30	26	Number of troops unknown to me.
1169	30	27	Convoy bears of us, as indicated next (per Compound Table, point of compass).
1170	30	28	I have made no captures.
1171	30	31	I have made a prize.
1172	30	32	I have made captures, *number* indicated next (per Compound Table).
1173	30	33	Vessel belonging to *place* indicated next, and whose name I shall spell, has been captured and taken to the *place* indicated by the second pair of signals following:—(both from Compound Table.)
1174	30	34	Vessel captured is a merchantman.
1175	30	35	Vessel captured is armed.
1176	30	36	I am chased.
1177	30	37	We are chased.
1178	30	38	I shall give chase.
1179	30	39	I have been chased by a man-of-war.
1180	30	40	I have been chased by a privateer.
1181	31	1	I have been chased by a suspicious sail.
1182	31	2	I must give up the chase.
1183	31	3	Give up the chase.
1184	31	4	Continue the chase.
1185	31	5	I don't know if vessel spoken of is armed or not.
1186	31	6	I have long guns only.
1187	31	7	I have short guns only.
1188	31	8	I have carronades only.
1189	31	9	I have both long guns, and carronades or short guns.
1190	31	10	My guns will not reach.

CHAPTER XI.

Relating to War, Fighting, &c.

Numerical order of Sentences.	Pair Signals. No.	No.	ANSWERS OR COMMUNICATIONS.
1191	31	11	My guns will reach.
1192	31	12	I have neither guns nor carronades.
1193	31	13	Vessel is within gun-shot.
1194	31	14	Vessel is not within gun-shot.
1195	31	15	Vessel or vessels, is or are line of battle ship, or line of battle ships.
1196	31	16	Vessel or vessels is or are a frigate or frigates.
1197	31	17	Vessel or vessels is or are a corvette (sloop of war) or corvettes.
1198	31	18	Vessel or vessels is or are a brig of war, or brigs of war.
1199	31	19	Vessel or vessels is or are a gun-boat, or gun-boats.
1200	31	20	Vessel is a man-of-war (or vessels are men-of-war).
1201	31	21	Vessel is a man-of-war-boat, or vessels are men-of-war-boats.
1202	31	22	Vessel is a packet, or vessels are packets.
1203	31	23	Vessel is a letter of marque, or vessels are letters of marque.
1204	31	24	Vessel is a privateer, or vessels are privateers.
1205	31	25	Firing is heard in quarter indicated next (per Compound Table, point of compass).
1206	31	26	Firing was heard in quarter indicated next (per Compound Table, point of compass).
1207	31	27	Peace is expected to take place between the *Powers* indicated next (per Compound Table).
1208	31	28	Vessel in sight, or bearing from me, as indicated next (per Compound Table, point of compass), is an enemy.
			Note.—If vessel in sight is to be understood, the last signal, when answered, is hauled nearly down, rehoisted, and struck.
1209	31	29	Vessels in sight, or bearing from me, as indicated next (per Compound Table, point of compass), are enemies.
			Note.—See preceding note.
1210	31	30	A great naval battle has been fought on the *date*, and off, or near, the *place*, as indicated by the two following pairs of signals:—(per Compound Table.)
1211	31	33	A great naval battle has been fought on the *date*, and in *latitude* and *longitude*, as indicated by the following pairs of signals:—(per Compound Table.)
1212	31	34	A naval battle of less consideration has been fought on the *date*, and off or near the *place*, as indicated by the two pairs of signals following:—(per Compound Table.)

CHAPTER XI.

Relating to War, Fighting, &c.

Numerical order of Sentences.	Pair Signals No.	Pair Signals No.	ANSWERS OR COMMUNICATIONS.
1213	31	35	A naval battle of less consideration has been fought on the *date*, and in *latitude* and *longitude*, as indicated by the pairs of signals following:—(per Compound Table.)
1214	31	36	A great land battle has been fought on the *date*, and near the *place*, as indicated by the two pairs of signals following:—(per Compound Table.)
1215	31	37	A land battle of less consideration has been fought on the *date*, and near the *place*, as indicated by the two pairs of signals following:—(per Compound Table.)
			Note.—Places not to be found in the Compound Table must be spelled.
1216	31	38	The battle was fought between *nations* indicated next (per Compound Table).
			Note.—The opposite parties, if more than one on each side, are distinguished by the signal for &, found in the same column as the names of countries, places, &c., in the Compound Table. See directions for the use of Compound Table.
1217	31	39	Battle began and lasted from *hour* to *hour* indicated next, by two pairs of signals (per Compound Table).
1218	31	40	Ships *sunk* and *taken*, as indicated next by two pairs of signals (per Compound Table, numbers).
			Note.—The first number signalized indicates vessels sunk, the second those that were taken. The signal for O is hoisted to indicate none, or the signal for knowledge is uncertain from Chap. XVII.
1219	32	1	No ships sunk or taken.
1220	32	2	The force of the *nation*, or the forces of the *nations*, indicated by signals following (per Compound Table), was or were beaten.
			Note.—If only one nation is to be signalized, the last signal of the pair indicating it, when answered, is hauled nearly down, rehoisted, and struck. But if more than one nation, this manœuvre is only to take place at the last signal of the pair indicating the last nation of the series.
1221	32	3	*Number* of killed, wounded, and taken prisoners on first-mentioned side in the preceding signal, computed as indicated by next pair signal (per Compound Table), and the *loss* on the other side will be indicated by the second pair of signals following:—(per Compound Table.)
1222	32	4	The commanding officer of *nation* indicated next (per Compound Table), is among the fallen.

CHAPTER XI.

Relating to War, Fighting, &c.

Numerical order of Sentences.	Pair Signals No.	No.	ANSWERS OR COMMUNICATIONS.
1223	32	5	An army or corps of *nation* indicated next (per Compound Table), has entered the territory of *nation* indicated by the second pair of signals following:—(per Compound Table.)
1224	32	6	An army or corps of the *nation* indicated next (per Compound Table), has joined an army or corps of the *nation* indicated by the second pair of signals following (per Compound Table), at the *place* to be indicated by the third pair of signals following (per Compound Table), or to be spelled.
1225	32	7	A corps commanded by the officer, whose name shall be spelt next, has joined another corps commanded by the officer whose name shall be spelt after that of the first, at the *place* indicated by the next pair signal (per Compound Table).
1226	32	8	The *place* or *fortress* indicated next (per Compound Table), or spelt, has surrendered by capitulation to the army of the *nation* indicated (per Compound Table).
1227	32	9	The *place* or *fortress* indicated by the next pair signal (per Compound Table), or by spelling, has been taken by storm by the army of the *nation* indicated (per Compound Table).
1228	32	10	A fleet or squadron of the *nation* indicated next (per Compound Table), has forced the entrance or strait at the *place* to be indicated by the second pair of signals following:—(per Compound Table.)
1229	32	11	The *place* indicated next (per Compound Table), has been taken by a naval expedition of *nation* indicated by the second pair of signals following:—(per Compound Table.)
1230	32	12	Fleet or squadron in the *port* indicated next (per Compound Table), is blockaded by fleet or squadron of *nation* to be indicated by the second pair of signals following:—(per Compound Table.)
1231	32	13	Fleet or squadron in the *port* indicated next (per Compound Table), is blockaded by fleets or squadrons of the *nations* to be indicated by the signals following:—(per Compound Table.) *Note.*—The last signal of the pair indicating the last nation of the series, when answered, is hauled nearly down, rehoisted, and struck.
1232	32	14	Succours are expected at *place* or *fortress* indicated next (per Compound Table, or by spelling).
1233	32	15	*Place* or *fortress* indicated next (per Compound Table, or by spelling), holds out still.

Numerical order of Sentences.	Pair Signals.		CHAPTER XI. *Relating to War, Fighting, &c.*
	No.	No.	ANSWERS OR COMMUNICATIONS.
1234	32	16	The army or corps of *nation* indicated next (per Compound Table), stands at present before the *place* indicated by second pair of signals (per Compound Table, or by spelling).
1235	32	17	The army or corps of *nation* indicated next (per Compound Table), is falling back.
1236	32	18	The armies of *nations* indicated next (per Compound Table), are standing over against each other.

CHAPTER XII.

Numerical order of Sentences.	PAIR SIGNALS. No.	No.	Relating to Reports of Vessels, Captain, Passengers and Crew, Ship News, &c. QUESTIONS.
1237	32	19	Where do you come from, and where are you bound to? *Ans.* From Compound Table, name of places in the same order as questions made.
1238	32	20	Where do you belong to, and what is the name of your vessel? *Ans.* From Compound Table, name of the place and the name of the vessel is spelled.
1239	32	21	What place do you come from last, and when did you leave it? *Ans.* Name of place and date, both from Compound Table.
1240	32	22	What date do you expect to be back? *Ans.* From Compound Table.
1241	32	23	What date do you think vessel belonging to *place* indicated next (per Compound Table), and whose name shall be spelt, may be expected back? *Ans.* Date from Compound Table.
1242	32	24	How many vessels may we expect from the place you are coming from? *Ans.* By number from Compound Table, or by an answer from Chap. XVII.
1243	32	25	What date did vessel belonging to *place* indicated next (per Compound Table), and whose name shall be spelt, leave harbour? *Ans.* Date from Compound Table.
1244	32	26	What date did vessel spoken of leave harbour? *Ans.* Date from Compound Table.
1245	32	27	Can you tell what date vessel belonging to *place* indicated next (per Compound Table), and whose name shall be spelt, left the *place* indicated by the second pair of signals following?—(per Compound Table.) *Ans.* Date from Compound Table, or from Chap. XVII.
1246	32	28	Do you know if vessel belonging to *place* indicated next (per Compound Table), and whose name shall be spelt, has arrived?

CHAPTER XII.

Relating to Reports of Vessels, Captain, Passengers and Crew, Ship News, &c.

Numerical order of Sentences.	Pair Signals. No.	Pair Signals. No.	QUESTIONS.
1247	32	29	Had vessel belonging to *place* indicated next (per Compound Table), and whose name shall be spelt, sailed when you left?
1248	32	30	Did more vessels leave port at the same time you did, with the same destination as you? *Ans.* By number from Compound Table, or from Chap. XVII.
1249	32	33	What vessel lies ready to sail for home, and where does it belong to? *Ans.* By spelling the name of the vessel, and a pair signal from Compound Table for place; or find an answer in Chap. XVII.
1250	32	34	What is the last arrival from *place* indicated next (per Compound Table)? *Ans.* By spelling the name of the vessel last arrived, and indicating place she belongs to (per Compound Table); or find an answer in Chap. XVII.
1251	32	35	What date did you arrive here? *Ans.* From Compound Table.
1252	32	36	What date do you intend to sail? *Ans.* From Compound Table.
1253	32	37	What date does vessel belonging to *place* indicated next (per Compound Table), and whose name I am going to spell, intend to sail? *Ans.* From Compound Table, or Chap. XVII.
1254	32	38	What course did vessel steer, when you saw her last? *Ans.* From Compound Table.
1255	32	39	Have you been in company with any other vessel on your present passage? *Ans.* By No; or an appropriate sentence from this chapter.
1256	32	40	What date did you part company with vessel in question? *Ans.* From Compound Table.
1257	33	1	Have you had any communication with any vessel on your present passage?
1258	33	2	Pass within hail.
1259	33	3	Where does vessel hailed belong to, and what is her name? *Ans.* From Compound Table for place, and by spelling the name.
1260	33	4	Where does vessel hailed come from, and where bound to? *Ans.* With pair signal for place coming from, and another for place bound to, both from Compound Table.

Numerical order of Sentences.	Pair Signals. No.	No.	CHAPTER XII. *Relating to Reports of Vessels, Captain, Passengers and Crew, Ship News, &c.*
			QUESTIONS.
1261	33	5	Have you spoken any vessel from *place* indicated next (per Compound Table), and were all well there?
1262	33	6	What do you take vessel, *bearing* as indicated next (per Compound Table), to be?
1263	33	7	Do you not take vessel, *bearing* from me as indicated next (per Compound Table), to be of a suspicious character?
1264	33	8	Are you well manned?
1265	33	9	Is the captain on board?
1266	33	10	What is the captain's name? *Ans.* By spelling the name.
1267	33	11	How does captain do?
1268	33	12	Will you (captain) come on board?
1269	33	13	Have you passengers on board, and how many? *Ans.* From Compound Table, numbers; or from Chap. XVII.
1270	33	14	Have you any lady passengers on board? *Ans.* By No, or by a sentence from this chapter.
1271	33	15	Can you accommodate some cabin passengers? *number* will be indicated next (per Compound Table).
1272	33	16	Can you accommodate lady passengers? *number* will be indicated next (per Compound Table).
1273	33	17	Can you give a passage to a child, of *so many* years old as indicated next (per Compound Table)?
1274	33	18	Can you receive as passengers, *so many* common men as will be indicated next (per Compound Table, numbers)? *Note.*—The four last sentences may be answered by *yes* or *no*; or by a sentence from this chapter, according to circumstances.
1275	33	19	Have you among your crew a man who can do the duty of a mate, and is willing to sail with me in that capacity, and will you take an able seaman in his place?
1276	33	20	Have you among your crew a man who understands cooking, and is willing to sail with me, and will you take another seaman from me in his place?
			ANSWERS OR COMMUNICATIONS.
1277	33	21	All are well on board vessel inquired after.
1278	33	22	Vessel inquired after has put into *place* indicated next (per Compound Table).
1279	33	23	Vessel inquired after has not yet been heard of.

Numerical order of Sentences.	Pair Signals No.	Pair Signals No.	CHAPTER XII. *Relating to Reports of Vessels, Captain, Passengers and Crew, Ship News, &c.* ANSWERS OR COMMUNICATIONS.
1280	33	24	Vessel belonging to *place* indicated next (per Compound Table), and whose name shall be spelled, will sail on *date* indicated by the second pair of signals following:—(per Compound Table.)
1281	33	25	Vessel in sight is beating to windward.
1282	33	26	Vessel belonging to *place* indicated next (per Compound Table), and whose name shall be spelled, bears of me as indicated by the second pair of signals following:—(per Compound Table, point of compass.)
1283	33	27	Vessel in sight, or vessel hailed, is a coaster.
1284	33	28	Vessel is an East Indiaman, or vessels are East Indiamen.
1285	33	29	Vessel is a merchantman, or vessels are merchantmen.
1286	33	30	Vessel is a transport, or vessels are transports.
			Note.—See Chap. XI. for vessels of war.
1287	33	31	I have lost sight of vessel, or vessels.
1288	33	32	A sail is in sight, bearing from me as indicated next (per Compound Table, point of compass).
			Note.—See Chap. XV. suspicious sails and pirates.
1289	33	35	I am well manned.
1290	33	36	I am weakly manned.
1291	33	37	I have been in company with vessel belonging to *place* indicated next (per Compound Table), and whose name shall be spelled.
1292	33	38	Captain is not on board.
1293	33	39	Captain will come on board.
			Note.—For more signals about captain, see Chap. V.
1294	33	40	I have a great many passengers, or I am full of passengers.
1295	34	1	I have no passenger.
1296	34	2	I can accommodate passengers, *number* will be indicated next (per Compound Table).
1297	34	3	I am to land passengers at *place* indicated next (per Compound Table).
1298	34	4	I have lady passengers on board.
			Note.—See more about passengers in Chap. V.
1299	34	5	I have been overhauled by a man-of-war of the *nation* indicated next (per Compound Table).
1300	34	6	I have been overhauled by a privateer of the *nation* indicated next (per Compound Table).

Numerical order of Sentences.	Pair Signals.		CHAPTER XII. *Relating to Reports of Vessels, Captain, Passengers and Crew, Ship News, &c.*
	No.	No.	ANSWERS OR COMMUNICATIONS.
1301	34	7	Vessel belonging to *place* indicated next (per Compound Table), and whose name I shall spell, has been overhauled by a man-of-war or privateer of the *nation* indicated by the second pair of signals following:—(per Compound Table.)
1302	34	8	Vessel was overhauled in *latitude* and *longitude* indicated next (per Compound Table).

CHAPTER XIII.

Numerical order of Sentences.	Pair Signals. No.	No.	*Relating to Articles of Trade and Cargo.* QUESTIONS.
			Note.—To avoid a needless repetition, all such articles as have already found a place in the list of stores contained in Chap. VIII. have been omitted here; it is therefore necessary to look back to the said eighth chapter for the names not to be found in this, and if any article should happen not to be found in either, recourse must be had to spelling.
1303	34	9	What does your cargo consist of? *Ans.* From this chapter, or from the list in Chap. VIII., or by spelling.
1304	34	10	Is your cargo destined for more places than one?
1305	34	11	What is the cargo of vessel belonging to *place* indicated next (per Compound Table), and whose name I am going to spell.
			SUNDRY ARTICLES.
1306	34	12	Ballast.
1307	34	13	Barilla.
1308	34	14	Bones.
1309	34	15	Bricks.
1310	34	16	Building stones.
1311	34	17	Cattle.
1312	34	18	Copper.
1313	34	19	Cotton twist.
1314	34	20	Dry goods.
1315	34	21	Dye wood.
1316	34	22	Earthenware.
1317	34	23	East India goods.
1318	34	24	Firestones, or firebricks.
1319	34	25	Flax-seed.
1320	34	26	Fruits.
1321	34	27	Furs.
1322	34	28	Glass-ware.
1323	34	29	Gold dust and ivory.
1324	34	30	Ironware.
1325	34	31	Hemp.
1326	34	32	Flax.
1327	34	35	Hides.

CHAPTER XIII.

Relating to Articles of Trade and Cargo.

Numerical order of Sentences.	Pair Signals. No.	No.	SUNDRY ARTICLES.
1328	34	36	Horses.
1329	34	37	Mules.
1330	34	38	Ice.
1331	34	39	Indigo.
1332	34	40	Wheat.
1333	35	1	Rye.
1334	35	2	Barley.
1335	35	3	Corn.
1336	35	4	Rags.
1337	35	5	Limestones.
1338	35	6	Timber.
1339	35	7	Lumber.
1340	35	8	Steel.
1341	35	9	Molasses.
1342	35	10	Rape-seed cakes (oil cakes).
1343	35	11	Sundries (or general cargo).
1344	35	12	Tobacco.
1345	35	13	Furniture.
1346	35	14	Naval stores.
1347	35	15	Provisions.
1348	35	16	Saltpetre.
1349	35	17	Sulphur, brimstone
1350	35	18	Slates.
1351	35	19	Tallow.
1352	35	20	Potash, pearl ashes.
1353	35	21	Wool (hair, bristles).
1354	35	22	Rape-seed.
1355	35	23	Pimento.
1356	35	24	Fustic.
1357	35	25	Logwood.
1358	35	26	Train oil, whale oil.
1359	35	27	Wine.
1360	35	28	Dried beef.
1361	35	29	Stock fish.
1362	35	30	Salt fish.
			ANSWERS OR COMMUNICATIONS.
1363	35	31	Part of my cargo is destined for *place* indicated next (per Compound Table), besides what is destined for place I am going to.

Numerical order of Sentences.	Pair Signals.		CHAPTER XIII. *Relating to Articles of Trade and Cargo.*
	No.	No.	ANSWERS OR COMMUNICATIONS.
1364	35	32	Part of my cargo is destined for different places besides that I am going to, *places* to be indicated by the following pair signals:—(per Compound Table.) *Note.*—The last signal of that pair which denotes the last of the places, is, when answered, to be hauled nearly down, rehoisted, and struck.
1365	35	33	My cargo is in good order. *Note.*—More about cargo is found in Chap. VII.

CHAPTER XIV.

Numerical order of Sentences.	Pair Signals No.	Pair Signals No.	Relating to Bars, Shoals, Rocks, and Reefs.
			QUESTIONS.
1366	35	34	Is the bar dangerous?
1367	35	37	May the bar be crossed?
1368	35	38	What water is there on the bar at half tide? *Ans.* By number of feet from Compound Table.
1369	35	39	Have I any thing to fear on the course I am steering?
			ANSWERS OR COMMUNICATIONS.
1370	35	40	Bar is dangerous.
1371	36	1	Bar is not dangerous.
1372	36	2	Bar cannot be crossed but at high water.
1373	36	3	Bar may be crossed at low water.
1374	36	4	You will be on shore or on the bank, if you do not alter your course.
1375	36	5	Do not stand too far on your present course, as reef or shoals stretch far out.
1376	36	6	Vessel *bearing* from me as indicated next (per Compound Table), will be on shore, or on the bank, if she continues.
1377	36	7	You may stand boldly in, keeping your lead going.
1378	36	8	There are rocks or shoals a-head of you.
1379	36	9	There are rocks or shoals on your starboard bow.
1380	36	10	There are rocks or shoals on your larboard bow.
1381	36	11	There are rocks or shoals *bearing* from me, as indicated next (per Compound Table).
1382	36	12	I see breakers *bearing* from me as indicated next (per Compound Table).
1383	36	13	Sea is looking shoal in the direction from me as indicated next (per Compound Table, point of compass).
1384	36	14	You are too near the rocks.
1385	36	15	Keep nearer to the rocks.
1386	36	16	A vessel is on the rocks, or on the reef or bank.
1387	36	17	Water is very shoal.
			Note.—For more sentences indicating danger, see " Selected Signals.'

CHAPTER XV.

Numerical order of Sentences.	Pair Signals. No.	No.	*Relating to Pirates and Suspicious Vessels.*
			QUESTIONS.
1388	36	18	Have you heard of pirates in these seas? *Note.*—"What do you take vessel to be?" is to be found under No. 1262, Chap. XII.; and "How is vessel armed," will be found in Chap. XI., No. 1138; likewise "Do you take vessel, bearing, as indicated next (from Compound Table) to be of a suspicious character?" is found under No. 1263, Chap. XII.
1389	36	19	Is there any thing uncommon in the appearance, or remarkable, about the hull and rigging of the pirate, by which she may be known?
			ANSWERS OR COMMUNICATIONS.
1390	36	20	I have heard nothing of pirates.
1391	36	21	A vessel has been plundered by a pirate in this neighbourhood.
1392	36	22	A vessel has been plundered by a pirate near the *place* indicated next (per Compound Table).
1393	36	23	I have been plundered, or a vessel has been plundered by a pirate in *latitude* and *longitude* as indicated next (per Compound Table). *Note.*—If "I have been plundered," is to be understood, the last signal of the sentence, when answered, is hauled nearly down, rehoisted, and struck.
1394	36	24	A vessel, supposed to be a pirate, has been seen, or is said to cruise, in *latitude* and *longitude* indicated next (per Compound Table).
1395	36	25	Vessel is a slaver.
1396	36	26	A slaver has been seen on the *date* in *latitude* and *longitude* as indicated (per Compound Table, date, latitude, and longitude in the same order).
1397	36	27	The slaver *steering* as indicated (per Compound Table).
1398	36	28	A pirate, or the pirate infesting the *seas* indicated next (per Compound Table) has been taken by a man-of-war. *Note.*—If the place is unknown, the last signal of the sentence, when answered, is hauled nearly down, rehoisted, and struck.
1399	36	29	Be upon your guard against the sail *bearing* from me, as indicated next (per Compound Table).
1400	36	30	Do not stand too close in shore in running along the coast or

Numerical order of Sentences.	Pair Signals. No.	Pair Signals. No.	CHAPTER XV. *Relating to Pirates and Suspicious Vessels.* ANSWERS OR COMMUNICATIONS.
			between the keys, or islands, as they are said to be suspicious, and a rendezvous for pirates.
1401	36	31	One of the vessels or boats in sight is suspicious.
1402	36	32	Vessels, or boats, in sight are suspicious.
1403	36	33	A suspicious sail hove in sight.
1404	36	34	A pirate is in sight *bearing* from me as indicated next (per Compound Table).
1405	36	37	Let us keep company until we have passed this place, or we have gained the offing, that we may be able to protect each other mutually.
1406	36	38	Pirate is, or is said to be, a ship.
1407	36	39	Pirate is, or is said to be, a brig.
1408	36	40	Pirate is, or is said to be, a hermaphrodite.
1409	37	1	Pirate is, or is said to be, a polacre.
1410	37	2	Pirate is, or is said to be, a lugger.
1411	37	3	Pirate is, or is said to be, a felucca.
1412	37	4	Pirate is, or is said to be, a schooner with two square topsails.
1413	37	5	Pirate is, or is said to be, a schooner with a square topsail.
1414	37	6	Pirate is, or is said to be, a schooner without square topsails.
1415	37	7	Pirate is, or is said to be, a sloop, or a one-masted vessel.
1416	37	8	Pirate is, or is said to be, a boat.
1417	37	9	Pirate is, or is said to be, a Barca (vessel used in the Mediterranean).
1418	37	10	Pirate is, or is said to be, a Proa (Malay vessel).
1419	37	11	Pirate is, or is said to be, armed with the *number* of guns indicated next (per Compound Table).
1420	37	12	Pirate has a gun a-midships.
1421	37	13	Pirate has no great guns nor carronades.
1422	37	14	Pirate is, or is said to be, strongly manned.
1423	37	15	Pirate is, or is said to be, weakly manned, *number* to be indicated next (per Compound Table).
			Note.—If the number is unknown, the last signal of the sentence, when answered, is hauled nearly down, rehoisted, and struck.
1424	37	16	Pirate has, or is said to have, *as many* ports in her side as indicated next (per Compound Table).
1425	37	17	Pirate has, or is said to have, no ports in her sides.
1426	37	18	Pirate has, or is said to have, a black hull.
1427	37	19	Pirate has, or is said to have, black yards and mast-heads.
1428	37	20	Pirate has, or is said to have, white yards, and mast-heads.
1429	37	21	Pirate has, or is said to have, black yards, and white mast-heads.

Numerical order of Sentences.	Pair Signals. No.	No.	CHAPTER XV. *Relating to Pirates, and Suspicious Vessels.* ANSWERS OR COMMUNICATIONS.
1430	37	22	Pirate is, or is said to be, black all over hull and spars.
1431	37	23	Pirate has, or is said to have, much rake in both her masts.
1432	37	24	Pirate's mainmast is said to rake a good deal more than her foremast.
1433	37	25	Pirate's foremast is said to rake forward.
1434	37	26	Pirate has, or is said to have, no figure head.
1435	37	27	Pirate has, or is said to have, a figure head.
1436	37	28	Pirate is said to have no topmast (if a single-masted vessel).
1437	37	29	Pirate is said to be of tonnage indicated next (per Compound Table, *number* of tons English).
1438	37	30	Pirate has an overhanging stern.
1439	37	31	Pirate has a narrow stern (pinktailed).
1440	37	32	Pirate is, or is said to be, copper-bottomed.
1441	37	33	Pirate is, or is said to be, not copper-bottomed.
1442	37	34	Pirate has other marks to be distinguished by, which I am going to spell.
1443	37	35	There is nothing remarkable in the appearance or about the hull or rigging of the pirate to distinguish her by.

CHAPTER XVI.

Numerical order of Sentences.	Pair Signals. No.	No.	*Containing Sentences of a Miscellaneous Nature.* QUESTIONS.
1444	37	36	Do you believe, or find it accurate?
1445	37	39	What do you advise me to do? is it advisable?
1446	37	40	Have you had an answer?
1447	38	1	Why don't you answer?
1448	38	2	Have you any thing for me?
1449	38	3	Have you any thing for the person whose name I shall spell?
1450	38	4	Is it satisfactory?
1451	38	5	Do you like it, or do you approve of it?
1452	38	6	Do they approve of it?
1453	38	7	Could not you inquire into it, or ascertain it?
1454	38	8	Why have you not inquired into it, or ascertained it?
1455	38	9	Have you any commands for the place I am going to?
1456	38	10	Shall we keep company?
1457	38	11	Have you any news?
1458	38	12	Will you dine with me to-day?
1459	38	13	Will you breakfast with me to-day?
1460	38	14	Will you take supper with me this evening?
1461	38	15	Will you allow me to take a friend with me?
1462	38	16	The person whose name shall be spelled, will not he come on board?
1463	38	17	Is fresh fish to be procured here? *Ans.* By Yes, or No.
1464	38	18	Are greens and fruits to be had here?
1465	38	19	Is fresh beef to be had here?
1466	38	20	Is wood for fuel to be had here?
1467	38	21	What is the matter on board of your vessel?
1468	38	22	What is the matter on board of vessel bearing from me, indicated next (per Compound Table, point of compass)?
1469	38	23	Where shall we rendezvous? *Ans.* With place, from Compound Table.
1470	38	24	At what time, or on what date, shall we be at rendezvous? *Ans.* Date, from Compound Table.
1471	38	25	Have you not understood my signal?
1472	38	26	Did you speak with him?
1473	38	27	Did you speak with her?
1474	38	28	Did you speak with them?
1475	38	29	Is it certain?

Numerical order of Sentences.	Pair Signals. No.	No.	CHAPTER XVI. *Containing Sentences of a Miscellaneous Nature.*
			QUESTIONS.
1476	38	30	Try to pick up something floating in my wake, though you should be obliged to yaw a little out of your course. *Note.*—If a man had fallen overboard, the signal for such an accident would, of course, have been made by hauling up and down the colours.
1477	38	31	Send off a wherry boat, or other shore boat.
1478	38	32	Send some fish off.
1479	38	33	Send off some fresh beef and greens.
1480	38	34	Send off a boat-load of wood.
1481	38	35	How many fish have you taken? *Ans.* From Compound Table.
1482	38	36	Has season been good for fishing?
1483	38	39	Are lodgings to be had on shore?
1484	38	40	Shall I make the attempt, or another attempt?
1485	39	1	Will you make an attempt?
1486	39	2	Have you made an attempt?
			ANSWERS OR COMMUNICATIONS.
1487	39	3	It is accurate.
1488	39	4	It is not accurate.
1489	39	5	I advise you to do as indicated next, by spelling.
1490	39	6	I have had no answer.
1491	39	7	I have had an answer.
1492	39	8	Send an answer.
1493	39	9	I will send an answer, or I shall answer.
1494	39	10	I have nothing for you, or person in question.
1495	39	11	It is not satisfactory.
1496	39	12	It is satisfactory.
1497	39	13	You do not pay attention.
1498	39	14	Vessel *bearing* from me, as indicated next (per Compound Table), does not pay attention.
1499	39	15	You may have all you want on shore.
1500	39	16	Nothing is to be had for money on shore.
1501	39	17	I thank you for your offer, but I know of nothing at present.
1502	39	18	There is nothing the matter with me, or the vessel indicated.
1503	39	19	I cannot trust my crew, keep near to me.
1504	39	20	There seems to be disturbance, or mutiny, broke out in vessel spoken of, or *bearing* from me as indicated next (per Compound Table). *Note.*—If vessel spoken of is to be understood, the last signal of the sentence, when answered, is hauled nearly down, rehoisted, and struck.
1505	39	21	Send a party of armed men to bring my crew to order.

Numerical order of Sentences.	Pair Signals. No.	No.	CHAPTER XVI. *Containing Sentences of a Miscellaneous Nature.* ANSWERS OR COMMUNICATIONS.
1506	39	22	Send a party of armed men on board of vessel, *bearing* from me as indicated next (per Compound Table), her crew being disorderly, or in mutiny.
1507	39	23	You ought to change your resolution.
1508	39	24	I do not wish to interfere with that business.
1509	39	25	I doubt its being practicable.
1510	39	26	I give my assent, or person concerned gives his or her assent.
1511	39	27	Let us part company.
1512	39	28	I can be detained no longer.
1513	39	29	I will detain you no longer.
1514	39	30	Report all well where you arrive.
1515	39	31	I left all well at the place I am coming from.
1516	39	32	I have had a loss in my family.
1517	39	33	Season has been good for fishing this year.
1518	39	34	Season has been middling for fishing this year.
1519	39	35	Season has been bad for fishing this year.
1520	39	36	There is plenty of fish to be caught here.
1521	39	37	I have been fortunate.
1522	39	38	I have not been fortunate.
1523	40	1	I shall pass you within hail.
1524	40	2	She is homeward bound.
1525	40	3	She is outward bound.
1526	40	4	I have no news.
1527	40	5	I had no news from vessel hailed or boarded.
1528	40	6	I shall make an attempt, or I shall make another attempt.
1529	40	7	Vessel belonging to *place* indicated next (per Compound Table), and whose name I shall spell, has made an attempt, but did not succeed.
1530	40	8	It is unnecessary.
1531	40	9	It is not probable.
1532	40	10	We dine, breakfast, or sup, at *hour* indicated next (per Compound Table).
1533	40	11	Take your friend or your passenger with you.
1534	40	12	You are welcome to bring with you whom you please.
1535	40	13	I shall take my friend or passenger with me.
1536	40	14	Vessel is making signals.
1537	40	15	I wish to speak with you.
1538	40	16	Spell the name or word.
1539	40	17	I have discovered, or I see people on shore, in the direction from me as indicated next (per Compound Table, point of compass).
1540	40	18	Vessel is a fishing-boat.

Numerical order of Sentences.	Pair Signals. No.	No.	CHAPTER XVI. *Containing Sentences of a Miscellaneous Nature.* ANSWERS OR COMMUNICATIONS.
1541	40	19	Vessels are fishing-boats.
1542	40	20	Vessel belongs to the country.
1543	40	21	Vessels belong to the country.
1544	40	22	Vessel is lateen rigged.
1545	40	23	Vessel is square rigged.
1546	40	24	*Place* indicated next (per Compound Table), is to be our rendezvous. *Note.*—This sentence is merely used as communicating, and not in answering the question concerning rendezvous to be found under No. 1469, first division of this chapter.
1547	40	25	I shall not have time to take leave of friends ashore, please to remember me to them, and make my excuses.
1548	40	26	Don't forget what you have promised me, or person whose name I shall spell. *Note.*—If "promised me" is to be understood, the last signal of the sentence, when answered, is hauled nearly down, rehoisted, and struck.

CHAPTER XVII.

Numerical order of Sentences.	Pair Signals. No.	Pair Signals. No.	*Containing General Answers.* ANSWERS OR COMMUNICATIONS.
1549	40	27	Signal made, (namely "the sentence indicated by it,") is not understood, although the flags were distinguished. *Note.*—This applies to both signals of the pair, if a pair signal; or if the signal happens to be taken from the selected signals, it means that one.
1550	40	28	I do not know, or my knowledge is uncertain.
1551	40	29	I have made no trial yet?
1552	40	30	I have made a fruitless trial.
1553	40	31	I have none, or I have had none.
1554	40	32	This case has not happened yet.
1555	40	33	I cannot comply with your request, or I am unable to do what your signal desires to be done.
1556	40	34	I must take time to consider, or I cannot answer immediately.
1557	40	35	Signal last made, annulled. *Note.*—This applies to both signals of the pair, if a pair signal; or if the last-made signal happen to be taken from the "selected signals," it simply means that one.
1558	40	36	One of my previous signals, which I am going to repeat next, is annulled. *Note.*—This may equally apply to a pair signal, and one of the "selected signals."
1559	40	37	Yes.
1560	40	38	No.

COMPOUND TABLE

OF THE

GREAT CODE OF SIGNALS.

DIRECTIONS

FOR THE USE OF THE FOLLOWING

COMPOUND TABLE.

This Table is composed of such objects of inquiry and conversation at sea, as generally require to be formed into so many different tables, with each its proper distinguishing signal; such as points of compass, date of the month, day of the week, hour of the day, degrees of latitude and longitude, names of harbours, towns, headlands, &c. The bringing of this heterogeneous matter together into one table, is founded upon the supposition, that a preceding sentence clearly shows the column of the Compound Table to which the signal made refers, or in which the answer to be given must be sought for. And though the signals for this table are the same which are used to express sentences, yet no mistake can happen, as it will always be clear from a preceding signal, whether the signal following, or the answer required, is to be looked for in the Compound Table, or not.

The nature and use of this table, simple as it is, may be best understood from a few examples.

The signal has been made, "What are the bearings, and distance in leagues, from the nearest land; and what is the name of it?"

This is answered from the Compound Table by pair signals, exactly in the same order as the questions follow each other in the above sentence, namely: first a pair of signals indicating the *point of compass;* next, another pair signal indicating the *number* of leagues; and lastly, a pair signal indicating the *name* of place, headland, &c., in question.

If the signal be, "What is the date of your latest newspaper?" the answer is given from the Compound Table by a pair of signals, indicating at once the *month,* and the *day* of the month.

If the signal be made, "How many days' quarantine may I expect to be put under?" it is answered by a pair signal from the Compound Table, indicating the *number* of days, which is found in the column for numbers in general in the said table.

And again, if the sentence signal has been made, "The battle has been fought between the nations indicated next (per Compound Table):" now suppose, for example, the British to have fought on one side, and the French and Spaniards on the other, then follows immediately after the above sentence, a pair signal from the Compound Table, indicating England; next another pair signal indicating *and,* or the sign & (which will be found in the same column with the nations, page 83), after that comes the pair signals indicating France, and lastly, the pair signal for Spain.

Note.—By pair signal is of course understood two successive signals, which necessary combination of two signals, to express a sentence, or an answer, or communication, from the Compound Table, forms the distinguishing feature of this Code of Signals.

By introducing this (&) in the Compound Table, we are enabled to distinguish between two opposite parties, where one or both are composed of two or more individuals. But where on both sides the parties consist of only one individual nation each, the sign & may be left out as superfluous.

If the signal be, "Battle began at hour indicated, and lasted till hour indicated," there follows, first a pair signal from Compound Table for the *hour* of the beginning, and next, another pair signal indicating the *hour* when battle ended.

And as a last example, the signal has been made, "What is your latitude and longitude to-day at noon?" This is answered by a pair signal for *latitude*, and another pair signal for *longitude*, both from Compound Table.

Note.—It ought to be observed that the Table only contains the latitude from O for every minute up to 11° 59′ separately for North and South latitude, as also the longitude (reckoned from the meridian of Greenwich East and West) is only to be found up to 11° 59′ separately for East and West longitude. Further, there has been added to this Table (see p. 121), four columns of *supplementary degrees*, beginning with 12°, and for every twelfth degree, the two first up to 84°, and the two last up to 168° inclusive. One of these columns belongs to latitude North, another to latitude South, the third to longitude East, and the fourth to longitude West. If the latitude, or longitude, to be signalized be greater than 11° 59′, it will either coincide with one of the supplementary numbers just mentioned, (in which case it may be signalized at once by the pair signal thereto belonging,) or it will fall between two such; in this case deduct the lowest of them from latitude or longitude given; signalize first the surplus, which now may be found in the Table with the number of minutes given, and next hoist the pair signal indicating the deducted supplementary number of degrees, which of course the party receiving the signal must add to the number of degrees and minutes before signalized.

It ought further to be observed, that where both latitude and longitude are desired, the latitude is first signalized in the above manner, and then follow the pair signals expressing the longitude. The following examples may serve to make this quite simple operation still clearer.

EXAMPLE I.

To be signalized, latitude 0° 0′, and long. E, 11° 37′.

1. Hoist pair signal No. 1 and No. 2, signifying 0° 0′ lat.
2. Hoist pair signal No. 18 and No. 12, signifying 11° 37′ long. E.

Note.—See note after this, marked *.

EXAMPLE II.

To be signalized, lat. 48° 0′ S, and long. E, 132° 0′.

1. Hoist pair signal No. 38, No. 11, signifying supplementary degrees 48° 0′ lat. S.
2. Hoist pair signal No. 39, No. 11, signifying supplementary degrees 132° 0′ long. E.

EXAMPLE III.

To be signalized lat. N, 12° 0′ and long. W, 19° 27′.

1. Hoist pair signal No. 38, No. 1, for supplementary degrees 12° 0′ lat. N.
2. Hoist pair signal No. 30, No. 31, for the surplus . . . 7° 27′ West long.
3. Hoist pair signal No. 39, No. 15, for supplementary degrees 12° 0′ West long.

EXAMPLE IV.

To be signalized lat. 24° 7′ S, and long. 0° 0′.

1. Hoist pair signal No. 19, No. 7, expressing surplus . . 0° 7′ lat. S.
2. Hoist pair signal No. 38, No. 9, for supplementary degrees 24° 0′ lat. S.
3. Hoist pair signal No. 1, No. 2, for . . . . 0° 0′ long.

EXAMPLE V.

To be signalized, lat. 71° 59′ N, and long. 179° 59′ E.

1. Hoist pair signal, No. 18, No. 36, for surplus . . . . . . 11 59′ lat. N.
2. Hoist pair signal, No. 38, No. 5, for supplementary degrees, 60° 0′ lat. N.
3. Hoist pair signal, No. 18, No. 36, for surplus . . . . . . 11° 59′ long. E.
4. Hoist pair signal, No. 39, No. 14, for supplementary degrees 168° 0′ long. E.

The supplementary degrees always form the closing part of a series of signals for latitude or longitude, where one or both happen to be greater than 11° 59′, but if no signal for degrees and minutes has been made previous to them, they themselves constitute the answer.*

It has been supposed in the preceding examples, that the latitude and the longitude were both to be communicated. The latter alone is however much oftener inquired about, and in that case, or should it be the latitude alone, it makes no difference in the manner of answering, as the number of degrees and minutes expressed by the answer will of course be understood to have reference to the subject mentioned in the question.

The Column for Numerals is made use of, not only to express any number of days, or distances in leagues, but the number of every thing else mentioned in the sentence previously signalized. But as the numbers do not reach higher than 1547, it is necessary to establish the following rules for expressing numbers of a greater amount:

1. A number above 1547, and up to 3094, is expressed by making the signal for 1547, and after that another pair signal for the rest; as for example, 1548 is expressed by making the pair signal for 1547, and after that the signal for 1. And 2501 is expressed by signalizing 1547, and next 954; and lastly, 3094, is expressed by signalizing 1547, and signalizing the same a second time.

2. Numbers higher than 3094 are expressed in the following manner: Suppose 3095 were to be signalized; then make the pair signal for 3, next another pair signal for 1000, and lastly, a pair signal for 95. If the number be 67,000; a pair signal for 67 is hoisted, and next, another for 1000. If the number be 67,025; signalize first 67, then 1000, and lastly, 25. And if 150,342; signalize 150, then 1000, and lastly, 342. In this manner a number as high as 1,547,000 may be expressed by hoisting the signal for 1547, and another for 1000; even 1,547,999, by a signal for 1547, another for 1000, and lastly, 999.

General Remark for Numbers. To indicate that no more numbers are to follow, whether there has been one, or more pair signals required, to express the number in question, the last signal of the closing pair-signal, when answered, is hauled nearly down, rehoisted, and struck.

The Column for Hours in this Table contains: first, the full hours of the day, and then follow hours and minutes together for every single minute in the hour, all the 24 hours round.

The Column for the Date, or Days of the month, contains: first, 31 days appropriated to express generally the date of the current month. Then follow the names of every day in the week, and after these follow all the months of the year, each having its proper number of days expressed singly.

The Column for Points of the Compass contains: first, the 32 points from

* When the latitude or longitude to be signalized is under 12 degrees, the communication can of course be made by one pair of signals; and in this case, to show that no supplementary degrees are to follow, the last signal of the pair is hauled nearly down, rehoisted, and struck.

North, through East, South, and West all round; and next follow every quarter-point from North all round the Compass.

The Column for Names of Nations, Countries, Seas, Towns, Headlands, &c., forms the last column of the Table, and here will be found inserted the sign &, immediately after the nations and countries.

That in certain seas and latitudes, where it is of peculiar interest to have your own reckoning and observations verified by comparing them with those of other vessels, as for example, in the Chops of the British Channel; or when making the Straits of Gibraltar; all over the North Sea, &c., one pair signal being sufficient to express the longitude to the minute, is an advantage of essential importance. The fundamental principle of this code of signals is *to save flags,* which made it necessary to adopt pair signals, or two successive signals, for expressing that, which other codes with a greater number of flags, are enabled to express by a single signal; it is nevertheless in this particular instance not at all inferior to them, as they require one signal for the degrees, and another for the minutes. The same is the case with the latitude from 0° to 11° 59′, North or South. The time may be expressed by one pair signal for every hour with the number of minutes given. Any given date may be expressed by a single pair signal, the month and day of the month at once, and the points of the compass may be signalized at once to a quarter point. In all the cases enumerated, the common codes require two or more signals to express what is here, by the assistance of the Compound Table, expressed by a pair signal; and distinguishing flags or pendants are also generally necessary to indicate the separate table to which these signals refer. The present system will, therefore, in many instances be found to have advantages over those hitherto in use, with respect to the number of signals required to express a particular subject; though, at first, the double signals required for every sentence may appear a disadvantage, from the loss of time and the additional trouble which they may seem to occasion.

It is hoped that the great advantage of this *Compound Table* will be obvious to the public, as by here making use of the same numbers of signals again, which have before been applied to the sentences, and that without any risk of confounding them together, we have been enabled to multiply the signals to several thousands, and thereby to express a variety of matter, and furnish a vast fund of conversation, such as has never been accomplished by other codes, possessed of far superior means, in twice the number of flags and pendants.

Note.—In consequence of the difference of spelling Names of Places in various Countries, the alphabetical arrangement cannot always be depended on beyond the first letter. In cases where the commencing letter in Danish is wholly different, the pair signals are placed against the English name, alphabetically arranged.

COMPOUND TABLE.

Pair Signals. No.	Pair Signals. No.	Numerals.	Time of the Day.	Latitude North.	Longitude East.	Days of the Month and Week.	Points of the Compass.	Nations, Countries, and the Sign &.
			H.	° ′	° ′			
1	2	0	Noon	0.00	0.00	1 Current	North	America, United States of
								Austria, see pair sigs. 38, 37
3	4	1	1 PM.	0. 1	0. 1	2 ,,	N by E	Bavaria
5	6	2	2 ,,	0. 2	0. 2	3 ,,	NNE	Belgium
7	8	3	3 ,,	0. 3	0. 3	4 ,,	NE by N	Bolivia
9	10	4	4 ,,	0. 4	0. 4	5 ,,	NE	Brazil
11	12	5	5 ,,	0. 5	0. 5	6 ,,	NE by E	Bremen
13	14	6	6 ,,	0. 6	0. 6	7 ,,	ENE	China
15	16	7	7 ,,	0. 7	0. 7	8 ,,	E by N	Columbia
17	18	8	8 ,,	0. 8	0. 8	9 ,,	E	Denmark
								Egypt, see pair signals 36, 35
19	20	9	9 ,,	0. 9	0. 9	10 ,,	E by S	England
21	22	10	10 ,,	0.10	0.10	11 ,,	ESE	France
23	24	11	11 ,,	0.11	0.11	12 ,,	SE by E	Greece
25	26	12	Midnight	0.12	0.12	13 ,,	SE	Hamburg
27	28	13	1 AM.	0.13	0.13	14 ,,	SE by S	Hanover
29	30	14	2 ,,	0.14	0.14	15 ,,	SSE	Holland
31	32	15	3 ,,	0.15	0.15	16 ,,	S by E	Japan
33	34	16	4 ,,	0.16	0.16	17 ,,	S	Lubec
35	36	17	5 ,,	0.17	0.17	18 ,,	S by W	Morocco
37	38	18	6 ,,	0.18	0.18	19 ,,	SSW	Mecklenburg-Schwerin
39	40	19	7 ,,	0.19	0.19	20 ,,	SW by S	Mexico
2	1	20	8 ,,	0.20	0.20	21 ,,	SW	Norway
4	3	21	9 ,,	0.21	0.21	22 ,,	SW by W	Oldenburg
6	5	22	10 ,,	0.22	0.22	23 ,,	WSW	Persia
8	7	23	11 ,,	0.23	0.23	24 ,,	W by S	Peru
			H. M.					
10	9	24	12. 1 ,,	0.24	0.24	25 ,,	W	Portugal
12	11	25	12. 2 ,,	0.25	0.25	26 ,,	W by N	Prussia
14	13	26	12. 3 ,,	0.26	0.26	27 ,,	WNW	Rome *(Ecclesiastical State)*
16	15	27	12. 4 ,,	0.27	0.27	28 ,,	NW by W	Russia
18	17	28	12. 5 ,,	0.28	0.28	29 ,,	NW	Saxony
20	19	29	12. 6 ,,	0.29	0.29	30 ,,	NW by N	Sardinia
22	21	30	12. 7 ,,	0.30	0.30	31 ,,	NNW	Sicilies
24	23	31	12. 8 ,,	0.31	0.31	——	N by W	Spain
26	25	32	12. 9 ,,	0.32	0.32	——	——	Sweden
28	27	33	12.10 ,,	0.33	0.33	Sunday	——	Tartary
30	29	34	12.11 ,,	0.34	0.34	Monday	——	Tuscany
32	31	35	12.12 ,,	0.35	0.35	Tuesday	——	Turkey
34	33	36	12.13 ,,	0.36	0.36	Wednesday	——	Wirtemberg
36	35	37	12.14 ,,	0.37	0.37	Thursday	——	Egypt
38	37	38	12.15 ,,	0.38	0.38	Friday	——	Austria
40	39	39	12.16 ,,	0.39	0.39	Saturday	——	&

COMPOUND TABLE.

Pair Signals. No	No.	Numerals.	Time of the Day.	Latitude North.	Longitude East.	Days of the Month.	Points of the Compass.	Nations, Countries, Seas, Seaports, Capes, &c.
			H. M.	° ′	° ′			
1	3	40	12.17 A.M.	0.40	0.40	1 Jan.	N ¼ E	Ahuus . . . Sweden
1	4	41	12.18 „	0.41	0.41	2 „	N ½ E	Aalborg . . . Jutland
1	5	42	12.19 „	0.42	0.42	3 „	N ¾ E	Aarhuus . . . do.
1	6	43	12.20 „	0.43	0.43	4 „	N by E ¼ E	Aaroe, *Light* . Little Belt
1	7	44	12.21 „	0.44	0.44	5 „	N by E ½ E	Abbeville . . France
1	8	45	12.22 „	0.45	0.45	6 „	N by E ¾ E	Aberdeen, *Light* . Scotland
1	9	46	12.23 „	0.46	0.46	7 „	NNE ¼ E	Abo . . . Finland
1	10	47	12.24 „	0.47	0.47	8 „	NNE ½ E	Abukir . . . Egypt
1	11	48	12.25 „	0.48	0.48	9 „	NNE ¾ E	Acapulco . America, W
1	12	49	12.26 „	0.49	0.49	10 „	NE ¾ N	Adam's Peak . . Ceylon
1	13	50	12.27 „	0.50	0.50	11 „	NE ½ N	Addon, *Island* . Indian Ocean
1	14	51	12.28 „	0.51	0.51	12 „	NE ¼ N	Admiralty Islands . Australasia
1	15	52	12.29 „	0.52	0.52	13 „	NE ¼ E	Adrianople . . Turkey
1	16	53	12.30 „	0.53	0.53	14 „	NE ½ E	Adriatic Sea . Italy & Dalmatia
1	17	54	12.31 „	0.54	0.54	15 „	NE ¾ E	Africa, *Coast of*
1	18	55	12.32 „	0.55	0.55	16 „	ENE ¾ N	African Islands . Indian Ocean
1	19	56	12.33 „	0.56	0.56	17 „	ENE ½ N	Agalega, *or* Gulega . do. do.
1	20	57	12.34 „	0.57	0.57	18 „	ENE ¼ N	Agersoe . . Great Belt
1	21	58	12.35 „	0.58	0.58	19 „	E by N ¾ N	Agnes, St., *Light* . Scilly Isles
1	22	59	12.36 „	0.59	0.59	20 „	E by N ½ N	Agosta . . . Sicily
1	23	60	12.37 „	1. 0	1. 0	21 „	E by N ¼ N	Agra . . . East Indies
1	24	61	12.38 „	1. 1	1. 1	22 „	E ¾ N	Aigues mortes, *Light* . France
1	25	62	12.39 „	1. 2	1. 2	23 „	E ½ N	d'Aiguillon, *Light* . do.
1	26	63	12.40 „	1. 3	1. 3	24 „	E ¼ N	d'Ailly, Cape, *Light* . do.
1	27	64	12.41 „	1. 4	1. 4	25 „	——	Aitkin's Rock . Ireland, NW
1	28	65	12.42 „	1. 5	1. 5	26 „	——	Aix, *Island, Light* . France, W
1	29	66	12.43 „	1. 6	1. 6	27 „	——	Ajaccio . . Corsica
1	30	67	12.44 „	1. 7	1. 7	28 „	——	Aland, *Isles* . Gulf of Bothnia
1	31	68	12.45 „	1. 8	1. 8	29 „	——	Alaschka . America, NW
1	32	69	12.46 „	1. 9	1. 9	30 „	——	Alban's Head, St. . England
1	33	70	12.47 „	1.10	1.10	31 „	——	Alboran, *Island* . Mediterranean
1	34	71	12.48 „	1.11	1.11	——	——	Aldabra, *Isles* . Indian Ocean
1	35	72	12.49 „	1.12	1.12	——	——	Alderney, *Isle* . English Channel
1	36	73	12.50 „	1.13	1.13	——	——	Aleppo . . . Syria
1	37	74	12 51 „	1.14	1.14	——	——	Aleutian *Islands*, Kamtschatka & America
1	38	75	12.52 „	1.15	1.15	——	——	Alexandria . United States of North America
1	39	76	12.53 „	1.16	1.16	——	——	Alexandria, *Light* . Egypt
1	40	77	12.54 „	1.17	1.17	——	——	Alfaques . . . Spain

COMPOUND TABLE.

Pair Signals. No	No.	Numerals.	Time of the Day.	Latitude North.	Longitude East.	Days of the Month.	Points of the Compass.	Nations, Countries, Seas, Seaports, Capes, &c.	
			H. M.	° ′	° ′				
2	3	78	12.55 AM.	1.18	1.18	1 Feb.	E ¼ S	ALGEZIRAS	Spain
2	4	79	12.56 „	1.19	1.19	2 „	E ½ S	ALGHIERI	Sardinia
2	5	80	12.57 „	1.20	1.20	3 „	E ¾ S	ALGIERS, *Light*	Africa, N
2	6	81	12.58 „	1.21	1.21	4 „	E by S ¼ S	ALICANT	Spain
2	7	82	12.59 „	1.22	1.22	5 „	E by S ½ S	ALMERIA	Spain
2	8	83	1. 1 „	1.23	1.23	6 „	E by S ¾ S	ALPHONSO, *Island*	Indian Ocean
2	9	84	1. 2 „	1.24	1.24	7 „	ESE ¼ S	D'ALPRECK, *Cape, Light*	France
2	10	85	1. 3 „	1.25	1.25	8 „	ESE ½ S	ALS, *Island*	Little Belt
2	11	86	1. 4 „	1.26	1.26	9 „	ESE ¾ S	ALCONA	Holstein
2	12	87	1. 5 „	1.27	1.27	10 „	SE ¾ E	ALTO VELA, *Island*	Hayti, S
2	13	88	1. 6 „	1.28	1.28	11 „	SE ½ E	AMACK, *Island*	Sealand
2	14	89	1. 7 „	1.29	1.29	12 „	SE ¼ E	AMBOYNA, *Island*	Molucca Isles
2	15	90	1. 8 „	1.30	1.30	13 „	SE ¼ S	AMELAND, *Island*	Holland
2	16	91	1. 9 „	1.31	1.31	14 „	SE ½ S	AMIRANTES *Islands*	Madagascar
2	17	92	1.10 „	1.32	1.32	15 „	SE ¾ S	AMRAM, *Island*	Denmark, W
2	18	93	1.11 „	1.33	1.33	16 „	SSE ¾ E	AMSTERDAM	Holland
2	19	94	1.12 „	1.34	1.34	17 „	SSE ½ E	AMSTERDAM, *Island*	Indian Ocean
2	20	95	1.13 „	1.35	1.35	18 „	SSE ¼ E	ANAMBAS, NORTH, *Island*	China Sea
2	21	96	1.14 „	1.36	1.36	19 „	S by E ¾ E	ANAMBAS, MIDDLE, *Island*	China Sea
2	22	97	1.15 „	1.37	1.37	20 „	S by E ½ E	ANAMBAS, SOUTH, *Island*	China Sea
2	23	98	1.16 „	1.38	1.38	21 „	S by E ¼ E	ANCONA, *Light*	Italy
2	24	99	1.17 „	1.39	1.39	22 „	S ¾ E	ANDAMAN, GREAT, *Isl.*	Bay of Bengal
2	25	100	1.18 „	1.40	1.40	23 „	S ½ E	ANDAMAN, LITTLE, *Isl.*	Bay of Bengal
2	26	101	1.19 „	1.41	1.41	24 „	S ¼ E	ANDREW, ST., *Cape*	Madagascar
2	27	102	1.20 „	1.42	1.42	25 „	———	ANDROS, *Island*	Archipelago
2	28	103	1.21 „	1.43	1.43	26 „	———	ANEGADE, *Island*	West Indies
2	29	104	1.22 „	1.44	1.44	27 „	———	ANGLESEA, *Island*	England
2	30	105	1.23 „	1.45	1.45	28 „	———	ANGOLA	Africa, W
2	31	106	1.24 „	1.46	1.46	29 „	———	ANGOSTURA	S. America
2	32	107	1.25 „	1.47	1.47	———	———	ANGUILLA, *Island*	West Indies
2	33	108	1.26 „	1.48	1.48	———	———	ANHOLT, *Island, Lights*	Cattegat
2	34	109	1.27 „	1.49	1.49	———	———	ANHOLT'S POINT	Cattegat
2	35	110	1.28 „	1.50	1.50	———	———	ANN, *Cape, Lights*	U. States, N.A.
2	36	111	1.29 „	1.51	1.51	———	———	ANNAPOLIS	Nova Scotia
2	37	112	1.30 „	1.52	1.52	———	———	ANNE, ST., *Cape*	Africa, W
2	38	113	1.31 „	1.53	1.53	———	———	ANTIBES	France
2	39	114	1.32 „	1.54	1.54	———	———	ANTICOSTI, *Island*	Labrador, S
2	40	115	1.33 „	1.55	1.55	———	———	D'ANTIFER, *Cape, Light*	France

COMPOUND TABLE.

Pair Signals. No	Pair Signals. No.	Numerals.	Time of the Day.	Latitude North.	Longitude East.	Days of the Month.	Points of the Compass.	Nations, Countries, Seas, Seaports, Capes, &c.
			H. M.	° ′	° ′			
3	1	116	1.34 A.M.	1.56	1.56	1 Mar.	S ¼ W	Antigua, *Island* . West Indies
3	2	117	1.35 ,,	1.57	1.57	2 ,,	S ½ W	Antiparos, *Island* . Archipelago
3	5	118	1.36 ,,	1.58	1.58	3 ,,	S ¾ W	Antongil Bay . Madagascar
3	6	119	1.37 ,,	1.59	1.59	4 ,,	SbyW ¼ W	Antonio, St., *Cape* Rio de la Plata
3	7	120	1.38 ,,	2. 0	2. 0	5 ,,	SbyW ½ W	Antonio, St., *Island* Cape Verd Isles
3	8	121	1.39 ,,	2. 1	2. 1	6 ,,	SbyW ¾ W	Antonio, St., *Cape* . Cuba, W
3	9	122	1.40 ,,	2. 2	2. 2	7 ,,	SSW ¼ W	Antwerp . . . Belgium
3	10	123	1.41 ,,	2. 3	2. 3	8 ,,	SSW ½ W	Apenrade . . Sleswick
3	11	124	1.42 ,,	2. 4	2. 4	9 ,,	SSW ¾ W	Arabia . . . . . Asia
3	12	125	1.43 ,,	2. 5	2. 5	10 ,,	SW ¾ S	Archangel . . . Russia
3	13	126	1.44 ,,	2. 6	2. 6	11 ,,	SW ½ S	Archangel, New . America, NW
3	14	127	1.45 ,,	2. 7	2. 7	12 ,,	SW ¼ S	Archipelago Morea & Asia Minor
3	15	128	1.46 ,,	2. 8	2. 8	13 ,,	SW ¼ W	Arcona, *Light* . . Rugen
3	16	129	1.47 ,,	2. 9	2. 9	14 ,,	SW ½ W	Arendal . . . Norway
3	17	130	1.48 ,,	2.10	2.10	15 ,,	SW ¾ W	Argos . . . . . Morea
3	18	131	1.49 ,,	2.11	2.11	16 ,,	WSW ¾ S	Arica . . Peru
3	19	132	1.50 ,,	2.12	2.12	17 ,,	WSW ½ S	Arcadia . . . Morea
3	20	133	1.51 ,,	2.13	2.13	18 ,,	WSW ¼ S	Arles . . . France
3	21	134	1.52 ,,	2.14	2.14	19 ,,	WbyS ¾ S	Arran, North, *Isl. Light* Ireland
3	22	135	1.53 ,,	2.15	2.15	20 ,,	WbyS ½ S	Arran, South, *Isl. Light* Ireland
3	23	136	1.54 ,,	2.16	2.16	21 ,,	WbyS ¼ S	Ascension, *Island* Atlantic Ocean
3	24	137	1.55 ,,	2.17	2.17	22 ,,	W ¾ S	Asoph, *Sea of* . . Russia
3	25	138	1.56 ,,	2.18	2.18	23 ,,	W ½ S	Aspoe Bake . Gulf of Finland
3	26	139	1.57 ,,	2.19	2.19	24 ,,	W ¼ S	Assumption, *Island* Indian Ocean
3	27	140	1.58 ,,	2.20	2.20	25 ,,	———	Astola, *Island* . . Persia
3	28	141	1.59 ,,	2.21	2.21	26 ,,	———	Astova, *Island* . Indian Ocean
3	29	142	2. 1 ,,	2.22	2.22	27 ,,	———	Astracan . . . Russia
3	30	143	2. 2 ,,	2.23	2.23	28 ,,	———	Athens . . . . . Morea
3	31	144	2. 3 ,,	2.24	2.24	29 ,,	———	Atschin . . . Sumatra
3	32	145	2. 4 ,,	2.25	2.25	30 ,,	———	Auckland's Islands . Australia
3	33	146	2. 5 ,,	2.26	2.26	31 ,,	———	Auculta, *Island* . Laccadives
3	34	147	2. 6 ,,	2.27	2.27	———	———	Augustine, St. . Madagascar
3	35	148	2. 7 ,,	2.28	2.28	———	———	Azores, *Islands* . Atlantic Ocean
3	36	149	2. 8 ,,	2.29	2.29	———	———	Baagoe, *Light* . Little Belt
3	37	150	2. 9 ,,	2.30	2.30	———	———	Babel-mandel, *Strait* . Red Sea
3	38	151	2.10 ,,	2.31	2.31	———	———	Bahama Islands . West Indies
3	39	152	2.11 ,,	2.32	2.32	———	———	Bahia, *Light* . . Brazil
3	40	153	2.12 ,,	2.33	2.33	———	———	Balaines, *Light* . . France

COMPOUND TABLE.

Pair Signals. No.	No.	Numerals.	Time of the Day.	Latitude North.	Longitude East.	Days of the Month.	Points of the Compass.	Nations, Countries, Seas, Seaports, Capes, &c.
			H. M.	° ′	° ′			
4	1	154	2.13 A.M.	2.34	2.34	1 Apr.	W ¼ N	Balambangan, *Island*. Borneo, N.
4	2	155	2.14 ,,	2.35	2.35	2 ,,	W ½ N	Ballasore . . Orissa Coast
								Baltic Sea, see pair sigs. 40, 26
4	5	156	2.15 ,,	2.36	2.36	3 ,,	W ¾ N	Baltimore . U. States, N.A.
4	6	157	2.16 ,,	2.37	2.37	4 ,,	W by N ¼ N	Banca, *Island* . Sumatra, E.
4	7	158	2.17 ,,	2.38	2.38	5 ,,	W by N ½ N	Banda, *Islands* . Moluccas
4	8	159	2.18 ,,	2.39	2.39	6 ,,	W by N ¾ N	Bantam . . . Java
4	9	160	2.19 ,,	2.40	2.40	7 ,,	WNW ¼ N	Bara, *Rock* . Orkneys, W.
4	10	161	2.20 ,,	2.41	2.41	8 ,,	WNW ½ N	Barbadoes, *Island* . West Indies
4	11	162	2.21 ,,	2.42	2.42	9 ,,	WNW ¾ N	Barbary, Coast . Africa, N
4	12	163	2.22 ,,	2.43	2.43	10 ,,	NW ¾ W	Barbuda, *Island* . West Indies
4	13	164	2.23 ,,	2.44	2.44	11 ,,	NW ½ W	Barcelona, *Light* . Spain
4	14	165	2.24 ,,	2.45	2.45	12 ,,	NW ¼ W	Barcelona . South America
4	15	166	2.25 ,,	2.46	2.46	13 ,,	NW ¼ N	Barfleur, *Light* . France
4	16	167	2.26 ,,	2.47	2.47	14 ,,	NW ½ N	Barnstaple, *Light* . U. States, N.A.
4	17	168	2.27 ,,	2.48	2.48	15 ,,	NW ¾ N	Barra, *Island* . Hebrides
4	18	169	2.28 ,,	2.49	2.49	16 ,,	NNW ¾ W	Barren, *Island* . Bay of Bengal
4	19	170	2.29 ,,	2.50	2.50	17 ,,	NNW ½ W	Barrow . America, N W
4	20	171	2.30 ,,	2.51	2.51	18 ,,	NNW ¼ W	Bartholomew, St., *Island* W. Indies
4	21	172	2.31 ,,	2.52	2.52	19 ,,	N by W ¾ W	Baskei, *or* Bashee Isles Manilla, N
4	22	173	2.32 ,,	2.53	2.53	20 ,,	N by W ½ W	Bassas, *Cape* . Africa, E
4	23	174	2.33 ,,	2.54	2.54	21 ,,	N by W ¼ W	Bass, *Strait*, N.S.W. Van Diemen's L.
4	24	175	2.34 ,,	2.55	2.55	22 ,,	N ¾ W	Bastia, *Light* . . Corsica
4	25	176	2.35 ,,	2.56	2.56	23 ,,	N ½ W	Basque Roads . . France
4	26	177	2.36 ,,	2.57	2.57	24 ,,	N ¼ W	Batavia . . . Java
4	27	178	2.37 ,,	2.58	2.58	25 ,,	———	Bath . . U. States, N. A.
4	28	179	2.38 ,,	2.59	2.59	26 ,,	———	Bayonne . . France
4	29	180	2.39 ,,	3. 0	3. 0	27 ,,	———	Beachy-Head, *Light* . England
4	30	181	2.40 ,,	3. 1	3. 1	28 ,,	———	Beaufort . U. States, N. A.
4	31	182	2.41 ,,	3. 2	3. 2	29 ,,	———	Belfast . . . Ireland
4	32	183	2.42 ,,	3. 3	3. 3	30 ,,	———	Belle-Isle, *Light* . France
4	33	184	2.43 ,,	3. 4	3. 4	———	———	Belle-Isle, Strait of . Newfoundland
4	34	185	2.44 ,,	3. 5	3. 5	———	———	Bell-Rock, *Light* . Scotland, E
4	35	186	2.45 ,,	3. 6	3. 6	———	———	Belt, Great Denmark
4	36	187	2.46 ,,	3. 7	3. 7	———	———	Belt, Little . Denmark
4	37	188	2.47 ,,	3. 8	3. 8	———	———	Bembridge, *Flag light* . Channel
4	38	189	2.48 ,,	3. 9	3. 9	———	———	Bencoolen . . Sumatra
4	39	190	2.49 ,,	3.10	3.10	———	———	Bengal . . East Indies
4	40	191	2.50 ,,	3.11	3.11	———	———	Bengal, Bay of . East Indies

COMPOUND TABLE.

Pair Signals. No.	Pair Signals. No.	Numerals.	Time of the Day. H. M.	Latitude North. ° ′	Longitude East. ° ′	Days of the Month.	Nations, Countries, Seas, Seaports, Capes, &c.	
5	1	192	2.51 AM.	3.12	3.12	1 May	Benguela	Africa, W.
5	2	193	2.52 ,,	3.13	3.13	2 ,,	Berbice	South America
5	3	194	2.53 ,,	3.14	3.14	3 ,,	Bergen	Norway
5	4	195	2.54 ,,	3.15	3.15	4 ,,	Behring's strait	Asia & America
5	7	196	2.55 ,,	3.16	3.16	5 ,,	Behring's, *or* Kamtschatka Sea	Kamtschatka
5	8	197	2.56 ,,	3.17	3.17	6 ,,	Berlingas, *Islands*	Portugal
5	9	198	2.57 ,,	3.18	3.18	7 ,,	Bermudas, *Islands*	Atlantic Ocean
5	10	199	2.58 ,,	3.19	3.19	8 ,,	Berwick, *Lights*	Scotland
5	11	200	2.59 ,,	3.20	3.20	9 ,,	Biarritz, *Light*	France
5	12	201	3. 1 ,,	3.21	3.21	10 ,,	Bilboa	Spain
5	13	202	3. 2 ,,	3.22	3.22	11 ,,	Billiton, *Island*	Sumatra & Borneo
5	14	203	3. 3 ,,	3.23	3.23	12 ,,	Bintang, *Island*	Straits of Malacca
5	15	204	3. 4 ,,	3.24	3.24	13 ,,	Biscay, Bay of	Spain & France
5	16	205	3. 5 ,,	3.25	3.25	14 ,,	Bissagos, *Islands*	Africa, W
5	17	206	3. 6 ,,	3.26	3.26	15 ,,	Bizerta	Africa, N
5	18	207	3. 7 ,,	3.27	3.27	16 ,,	Biorneborg	Russia
5	19	208	3. 8 ,,	3.28	3.28	17 ,,	Black Rocks	East Indies
							Black Sea	see pair signals 34, 29
5	20	209	3. 9 ,,	3.29	3.29	18 ,,	Blacdale's Coral Reef	Rio Maranon, E
5	21	210	3.10 ,,	3.30	3.30	19 ,,	Blanco, *Cape*	Africa, W
5	22	211	3.11 ,,	3.31	3.31	20 ,,	Blanquilla, *Island*	Caribbean Sea
5	23	212	3.12 ,,	3.32	3.32	21 ,,	Block Island	United States, N. A.
5	24	213	3.13 ,,	3.33	3.33	22 ,,	Blokkenbierg	Denmark, W
5	25	214	3.14 ,,	3.34	3.34	23 ,,	Blossom, *Cape*	America, NW
5	26	215	3.15 ,,	3.35	3.35	24 ,,	Blyth, *Lights*	England
5	27	216	3.16 ,,	3.36	3.36	25 ,,	Bodoe	Nordland
5	28	217	3.17 ,,	3.37	3.37	26 ,,	Bolsaxen, *Bank*	Great Belt
5	29	218	3.18 ,,	3.38	3.38	27 ,,	Bolt Head	England
5	30	219	3.19 ,,	3.39	3.39	28 ,,	Bombay	Hindostan
5	31	220	3.20 ,,	3.40	3.40	29 ,,	Bon, *Cape*	Africa, N
5	32	221	3.21 ,,	3.41	3.41	30 ,,	Bona	Africa, N
5	33	222	3.22 ,,	3.42	3.42	31 ,,	Bona vista, *Cape*	Newfoundland
5	34	223	3.23 ,,	3.43	3.43	——	Bona vista, *Island*	Cape Verd Isles
5	35	224	3.24 ,,	3.44	3.44	——	Bonifacio	Corsica
5	36	225	3.25 ,,	3.45	3.45	——	Bordeaux, *Light*	France
5	37	226	3.26 ,,	3.46	3.46	——	Borgo	Russia
5	38	227	3.27 ,,	3.47	3.47	——	Borkum, *Island, Light*	Hanover
5	39	228	3.28 ,,	3.48	3.48	——	Borneo, *Island*	Sunda Islands
5	40	229	3.29 ,,	3.49	3.49	——	Bornholm, *Island*	Baltic

COMPOUND TABLE.

Pair No.	Signals No.	Numerals.	Time of the Day.	Latitude North.	Longitude East.	Days of the Month.	Nations, Countries, Seas, Seaports, Capes, &c.	
			H. M.	° ′	° ′			
6	1	230	3.30 A.M.	3.50	3.50	1 June	Boston, *Light*	United States, N.A.
6	2	231	3.31 „	3.51	3.51	2 „	Botany Bay	New Holland
6	3	232	3.32 „	3.52	3.52	3 „	Bothnia, Gulf of	Russia & Sweden
6	4	233	3.33 „	3.53	3.53	4 „	Boulogne, *Lights*	France
6	7	234	3.34 „	3.54	3.54	5 „	Bourbon, Isle of	Indian Ocean
6	8	235	3.35 „	3.55	3.55	6 „	Bovbierg	Denmark, W
6	9	236	3.36 „	3.56	3.56	7 „	Boyador, *Cape*	Africa, W
6	10	237	3.37 „	3.57	3.57	8 „	Brahestad	Finland
6	11	238	3.38 „	3.58	3.58	9 „	Brala, *Pulo*	Chinese Sea
6	12	239	3.39 „	3.59	3.59	10 „	Brazil Rock	Ireland, W
6	13	240	3.40 „	4. 0	4. 0	11 „	Brasse, *Pulo*	Sumatra
6	14	241	3.41 „	4. 1	4. 1	12 „	Brava, *Island*	Cape Verd Isles
6	15	242	3.42 „	4. 2	4. 2	13 „	Bredgrund	The Sound
6	16	243	3.43 „	4. 3	4. 3	14 „	Brehat, Heaux de, *Light*	France
6	17	244	3.44 „	4. 4	4. 4	15 „	Brest	France
6	18	245	3.45 „	4. 5	4. 5	16 „	Breton, *Cape, Island*	North America
6	19	246	3.46 „	4. 6	4. 6	17 „	Bridgewater, *Light*	England
6	20	247	3.47 „	4. 7	4. 7	18 „	Bridport	England
6	21	248	3.48 „	4. 8	4. 8	19 „	Brighton, *Light*	England
6	22	249	3.49 „	4. 9	4. 9	20 „	Bristol	England
6	23	250	3.50 „	4.10	4.10	21 „	Broad Fourteens Bank	North Sea
6	24	251	3.51 „	4.11	4 11	22 „	Brusterort, *Lights*	Prussia
6	25	252	3.52 „	4.12	4.12	23 „	Buchanness, *Light*	Scotland
6	26	253	3.53 „	4.13	4.13	24 „	Buen Ayre, *Island*	Caribbean Sea
6	27	254	3.54 „	4.14	4.14	25 „	Buenos-Ayres	Rio de la Plata
6	28	255	3.55 „	4.15	4.15	26 „	Bugia	Africa, N
6	29	256	3.56 „	4.16	4.16	27 „	Bugio, *Light*	Portugal
6	30	257	3.57 „	4.17	4.17	28 „	Bulak	Egypt
6	31	258	3.58 „	4.18	4.18	29 „	Bulk, *Light*	Sleswick
6	32	259	3.59 „	4.19	4.19	30 „	Busum	Holstein
6	33	260	4. 1 „	4.20	4.20	——	Butt of Lewis	Isle of Lewis
6	34	261	4. 2 „	4.21	4.21	——	Buttonness, *Lights*	Scotland
6	35	262	4. 3 „	4.22	4.22	——	Button's Island	Labrador
6	36	263	4. 4 „	4.23	4.23	——	Butrinto	Albania
6	37	264	4. 5 „	4.24	4.24	——	Byron's Island	Australia
6	38	265	4. 6 „	4.25	4.25	——	Cabrera, *Island*	Majorca, S
6	39	266	4. 7 „	4.26	4.26	——	Cadiz	Spain
6	40	267	4. 8 „	4.27	4.27	——	Cagliari	Sardinia

COMPOUND TABLE.

Pair Signals. No.	Pair Signals. No.	Numerals.	Time of the Day.	Latitude North.	Longitude East.	Days of the Month.	Nations, Countries, Seas, Seaports, Capes, &c.	
			H. M.	° ′	° ′			
7	1	268	4. 9 A.M.	4.28	4.28	1 July	Calais, *Light*	France
7	2	269	4.10 „	4.29	4.29	2 „	Calcutta	Hindostan
7	3	270	4.11 „	4.30	4.30	3 „	Calicut	Malabar Coast
7	4	271	4.12 „	4.31	4.31	4 „	California	America, W
7	5	272	4.13 „	4.32	4.32	5 „	Callao	Peru
7	6	273	4.14 „	4.33	4.33	6 „	Calvi	Corsica
7	9	274	4.15 „	4.34	4.34	7 „	Calymere Point	Coromandel Coast
7	10	275	4.16 „	4.35	4.35	8 „	Camargue, *Light*	France
7	11	276	4.17 „	4.36	4.36	9 „	Campeche	Yucatan
7	12	277	4.18 „	4.37	4.37	10 „	Canada	North America
7	13	278	4.19 „	4.38	4.38	11 „	Channel, English	England & France
7	14	279	4.20 „	4.39	4.39	12 „	——— or Strait between Calais & Dover	
7	15	280	4.21 „	4.40	4.40	13 „	Canaria, *Island*	Canary Isles
7	16	281	4.22 „	4.41	4.41	14 „	Canary Islands	Atlantic Ocean
7	17	282	4.23 „	4.42	4.42	15 „	Candia, *Island*	Mediterranean
7	18	283	4.24 „	4.43	4.43	16 „	Canton	China
7	19	284	4.25 „	4.44	4.44	17 „	Cape of Good Hope, *Light*	Africa, S
7	20	285	4.26 „	4.45	4.45	18 „	Capri, *Island*	Italy
7	21	286	4.27 „	4.46	4.46	19 „	Carbonara, *Cape*	Sardinia
7	22	287	4.28 „	4.47	4.47	20 „	Cargados Garajos, *Reef*	Indian Ocean
7	23	288	4.29 „	4.48	4.48	21 „	Carimon Java, *Island*	Java Sea
7	24	289	4.30 „	4.49	4.49	22 „	Carlebye, Old	Finland
7	25	290	4.31 „	4.50	4.50	23 „	Carlebye, New	Finland
7	26	291	4.32 „	4.51	4.51	24 „	Carlisle	England
7	27	292	4.33 „	4.52	4.52	25 „	Carlshamn	Sweden
7	28	293	4.34 „	4.53	4.53	26 „	Carlscrona	Sweden
7	29	294	4.35 „	4.54	4.54	27 „	Carlsoe, Great	Gothland, W
7	30	295	4.36 „	4.55	4.55	28 „	Carloe	Gulf of Bothnia
7	31	296	4.37 „	4.56	4.56	29 „	Carnom, *Pulo*	Gulf of Siam
7	32	297	4.38 „	4.57	4.57	30 „	Cardigan Bay	England
7	33	298	4.39 „	4.58	4.58	31 „	Caroline Islands	Australia
7	34	299	4.40 „	4.59	4.59	———	Carpentaria	New Holland
7	35	300	4.41 „	5. 0	5. 0	———	Carrickfergus	Ireland
7	36	301	4.42 „	5. 1	5. 1	———	Carthagena	Spain
7	37	302	4.43 „	5. 2	5. 2	———	Carthagena	South America
7	38	303	4.44 „	5. 3	5. 3	———	Carvoeira, *Cape, Light*	Portugal
7	39	304	4.45 „	5. 4	5. 4	———	Casco Bay	North America
7	40	305	4.46 „	5. 5	5. 5	———	Caspian Sea	Asia

COMPOUND TABLE.

Pair Signals. No.	No.	Numerals	Time of the Day.	Latitude North.	Longitude East.	Days of the Month.	Nations, Countries, Seas, Seaports, Capes, &c.
			H.M.	° ′	° ′		
8	1	306	4.47 A.M.	5. 6	5. 6	1 Aug.	Castine . . United States, N.A.
8	2	307	4.48 „	5. 7	5. 7	2 „	Casquet, *Rock, Lights* . Channel
8	3	308	4.49 „	5. 8	5. 8	3 „	Catania . . . Sicily
8	4	309	4.50 „	5. 9	5. 9	4 „	Catharinendal, *Light* . Russia
8	5	310	4.51 „	5.10	5.10	5 „	Cattaro . . . Dalmatia
							Cattegat, see Kattegal
8	6	311	4.52 „	5.11	5.11	6 „	Cavalier, *Cape* . . Asia Minor
8	9	312	4.53 „	5.12	5.12	7 „	Cavallo, *Cape, Light* . . Naples
8	10	313	4.54 „	5.13	5.13	8 „	Caxo, *Island* . . Mediterranean
8	11	314	4.55 „	5.14	5.14	9 „	Caycos Islands . West Indies
8	12	315	4.56 „	5.15	5.15	10 „	Cayenne, *French* . South America
8	13	316	4.57 „	5.16	5.16	11 „	Cayenne, *English* . South America
8	14	317	4.58 „	5.17	5.17	12 „	Cayeux, *Light* . . France
8	15	318	4.59 „	5.18	5.18	13 „	Cephalonica, *Island* . Ionian Islands
8	16	319	5. 1 „	5.19	5.19	14 „	Cesir do Mar, *Pulo* . China Sea
8	17	320	5. 2 „	5.20	5.20	15 „	Celebes, *Island* . . Isles of Sunda
8	18	321	5. 3 „	5.21	5.21	16 „	Ceram, *Island* . . Molucca Islands
8	19	322	5. 4 „	5.22	5.22	17 „	Cerigo, *Island* . . Morea, S
8	20	323	5. 5 „	5.23	5.23	18 „	Cette, *Lights* . France
8	21	324	5. 6 „	5.24	5.24	19 „	Ceuta, *Cape* . . Africa, N
8	22	325	5. 7 „	5.25	5.25	20 „	Ceylon, *Island* . . Hindostan, S
8	23	326	5. 8 „	5.26	5.26	21 „	Chagos, *Islands* . Indian Ocean
8	24	327	5. 9 „	5.27	5.27	22 „	Chagre . South America
8	25	328	5.10 „	5.28	5.28	23 „	Channel, North . . Ireland
8	26	329	5.11 „	5.29	5.29	24 „	Charente . . France
8	27	330	5.12 „	5.30	5.30	25 „	Charles, *Cape* . United States, N.A.
8	28	331	5.13 „	5.31	5.31	26 „	Charlestown . United States, N.A.
8	29	332	5.14 „	5.32	5.32	27 „	Q. Charlotte's Islands America, NW
8	30	333	5.15 „	5.33	5.33	28 „	Chassiron, *Light* . France
8	31	334	5.16 „	5.34	5.34	29 „	Chatham . . . England
8	32	335	5.17 „	5.35	5.35	30 „	Chatham, *Island* . New Zealand, E
8	33	336	5.18 „	5.36	5.36	31 „	Cheduba, *Island* . Bay of Bengal
8	34	337	5.19 „	5.37	5.37	——	Chelidoni, *Cape* . Asia Minor
8	35	338	5.20 „	5.38	5.38	——	Cherbourg . . France
8	36	339	5.21 „	5.39	5.39	——	Chesapeak . United States, N.A.
8	37	340	5.22 „	5.40	5.40	——	Chewfgereen . America, NW
8	38	341	5.23 „	5.41	5.41	——	Chiavari . . . Italy
8	39	342	5.24 „	5.42	5.42	——	Chili . . South America
8	40	343	5.25 „	5.43	5.43	——	China Sea . . China

COMPOUND TABLE.

Pair Signals. No.	Pair Signals. No.	Numerals.	Time of the Day. H.M.	Latitude North. ° ′	Longitude East. ° ′	Days of the Month.	Nations, Countries, Seas, Seaports, Capes, &c.	
9	1	344	5.26 AM.	5.44	5.44	1 Sept.	Christiania .	. Norway
9	2	345	5.27 ,,	5.45	5.45	2 ,,	Christiania Firth .	Norway
9	3	346	5.28 ,,	5.46	5.46	3 ,,	Christianopel .	. Sweden
9	4	347	5.29 ,,	5.47	5.47	4 ,,	Christiansand .	. Norway
9	5	348	5.30 ,,	5.48	5.48	5 ,,	Christiansborg .	. Africa, W
9	6	349	5.31 ,,	5.49	5.49	6 ,,	Christianstæd .	. St. Croix
9	7	350	5.32 ,,	5.50	5.50	7 ,,	Christiansund .	. Norway
9	8	351	5.33 ,,	5.51	5.51	8 ,,	Christiansoe, *Light* .	Baltic
9	11	352	5.34 ,,	5.52	5.52	9 ,,	Christinestad .	. Finland
9	12	353	5.35 ,,	5.53	5.53	10 ,,	Christmas Island .	Indian Ocean
9	13	354	5.36 ,,	5.54	5.54	11 ,,	Christophers, St., or St. Kitts, *Isl.*	W. Ind.
9	14	355	5.37 ,,	5.55	5.55	12 ,,	Cimbrishamn .	Sweden
9	15	356	5.38 ,,	5.56	5.56	13 ,,	Cintra Bay .	. Africa, W
9	16	357	5.39 ,,	5.57	5.57	14 ,,	Civita Vecchia, *Light*	. Italy
9	17	358	5.40 ,,	5.58	5.58	15 ,,	Clare, *Light* .	. Ireland
9	18	359	5.41 ,,	5.59	5.59	16 ,,	Clarence, Port .	America, NW
9	19	360	5.42 ,,	6. 0	6. 0	17 ,,	Clear, *Cape, Light* .	Ireland
9	20	361	5.43 ,,	6. 1	6. 1	18 ,,	Clyde .	. Scotland
9	21	362	5.44 ,,	6. 2	6. 2	19 ,,	Cape Coast .	. Africa, W
9	22	363	5.45 ,,	6. 3	6. 3	20 ,,	Cochin .	. Malabar Coast
9	23	364	5.46 ,,	6. 4	6. 4	21 ,,	Coco's Islands .	Indian Ocean
9	24	365	5.47 ,,	6. 5	6. 5	22 ,,	Coco Island, North	. Indian Ocean
9	25	366	5.48 ,,	6. 6	6. 6	23 ,,	Cod, *Cape, Light*	United States, N.A.
9	26	367	5.49 ,,	6. 7	6. 7	24 ,,	Codgone Bay .	Golconda Coast
9	27	368	5.50 ,,	6. 8	6. 8	25 ,,	Coetivy, *Island* .	Indian Ocean
9	28	369	5.51 ,,	6. 9	6. 9	26 ,,	Colberg .	. Pomerania
9	29	370	5.52 ,,	6.10	6.10	27 ,,	Colombo .	. Ceylon
9	30	371	5.53 ,,	6.11	6.11	28 ,,	Comacchio .	. Italy
9	31	372	5.54 ,,	6.12	6.12	29 ,,	Commerce, Tour du, *Light*	. France
9	32	373	5.55 ,,	6.13	6.13	30 ,,	Comorin, *Cape* .	Malabar Coast
9	33	374	5.56 ,,	6.14	6.14	——	Comoro, *Island*	Mosambique Channel
9	34	375	5.57 ,,	6.15	6.15	——	Conception .	South America
9	35	376	5.58 ,,	6.16	6.16	——	Congo .	. Africa, W
9	36	377	5.59 ,,	6.17	6.17	——	Coningbeg, *Float-Light*	St. George's Cha.
9	37	378	6. 1 ,,	6.18	6.18	——	Connecticut .	United States, N.A.
9	38	379	6. 2 ,,	6.19	6.19	——	Constantinople .	. Turkey
9	39	380	6. 3 ,,	6.20	6.20	——	Cook's Straits .	New Zealand
9	40	381	6. 4 ,,	6.21	6.21	——	Cook's Islands .	Australia
							Copenhagen, see Kiobenhavn	

COMPOUND TABLE.

Pair Signals. No.	No.	Numerals.	Time of the Day.	Latitude North.	Longitude East.	Days of the Month.	Nations, Countries, Seas, Seaports, Capes, &c.	
			h. m.	° '	° '			
10	1	382	6. 5 am.	6.22	6.22	1 Oct.	Cordouan Island, *Light* .	France, W
10	2	383	6. 6 „	6.23	6.23	2 „	Corea, Peninsula of . .	China
10	3	384	6. 7 „	6.24	6.24	3 „	Corfu, *Island, Light* .	Ionian Isles
10	4	385	6. 8 „	6.25	6.25	4 „	Cork, *Light* . . .	Ireland
10	5	386	6. 9 „	6.26	6.26	5 „	Coromandel Coast .	East Indies
10	6	387	6.10 „	6.27	6.27	6 „	Coron	Morea
10	7	388	6.11 „	6.28	6.28	7 „	Corrientes, *Cape* . .	Columbia
10	8	389	6.12 „	6.29	6.29	8 „	———— *Cape*	Africa, E
10	11	390	6.13 „	6.30	6.30	9 „	Corsica, *Island* . .	Mediterranean
10	12	391	6.14 „	6.31	6.31	10 „	Corsoer . . .	Sealand
10	13	392	6.15 „	6.32	6.32	11 „	Corunna, *Light* . . .	Spain
10	14	393	6.16 „	6.33	6.33	12 „	Corvo, *Island* . . .	Azores
10	15	394	6.17 „	6.34	6.34	13 „	Cowes . .	Isle of Wight
10	16	395	6.18 „	6.35	6.35	14 „	Crab Island . .	West Indies
10	17	396	6.19 „	6.36	6.36	15 „	Creux, *Cape, Light* . . .	Spain
10	18	397	6.20 „	6.37	6.37	16 „	Croix, St., *Island* .	West Indies
10	19	398	6.21 „	6.38	6.38	17 „	Cromer, *Light* . . .	England
10	20	399	6.22 „	6.39	6.39	18 „	Cronstadt	Russia
10	21	400	6.23 „	6.40	6.40	19 „	Crozets, or Desert Islands	Indian Ocean
10	22	401	6.24 „	6.41	6.41	20 „	Cruz, Santa, *Islands* . .	Australia
10	23	402	6.25 „	6.42	6.42	21 „	Cuba, *Island* . . .	West Indies
10	24	403	6.26 „	6.43	6.43	22 „	Culebra	America, W
10	25	404	6.27 „	6.44	6.44	23 „	Cumana . . .	South America
10	26	405	6.28 „	6.45	6.45	24 „	Cumberland Straits	Baffin's & Hudson's Bay
10	27	406	6.29 „	6.46	6.46	25 „	Curaçao, *Island* .	South America
10	28	407	6.30 „	6.47	6.47	26 „	Curia Muria, *Island* . .	Arabia
10	29	408	6.31 „	6.48	6.48	27 „	Cutch, Gulf of . .	Guzerat, N
10	30	409	6.32 „	6.49	6.49	28 „	Cuxhaven, *Light* . .	Hanover
10	31	410	6.33 „	6.50	6.50	29 „	Cyprus, *Island* . .	Mediterranean
10	32	411	6.34 „	6.51	6.51	30 „	Dagerort, *Light* . . .	Dagoe
10	33	412	6.35 „	6.52	6.52	31 „	Dagoe	Baltic
10	34	413	6.36 „	6.53	6.53	———	Dalmatia	Adriatic
10	35	414	6.37 „	6.54	6.54	———	Damietta	Egypt
10	36	415	6.38 „	6.55	6.55	———	Dampier's Archipelago	New Holland
10	37	416	6.39 „	6.56	6.56	———	Dantzic, *Light* . . .	Prussia
10	38	417	6.40 „	6.57	6.57	———	Dardanelles	Turkey
10	39	418	6.41 „	6.58	6.58	———	Darien . . .	South America
10	40	419	6.42 „	6.59	6.59	———	Darserort . . .	Mecklenburg

COMPOUND TABLE.

Pair Signals. No.	Pair Signals. No.	Numerals.	Time of the Day.	Latitude North.	Longitude East.	Days of the Month.	Nations, Countries, Seas, Seaports, Capes, &c.	
			H. M.	° ′	° ′			
11	1	420	6.43 AM.	7. 0	7. 0	1 Nov.	Dartmouth	England
11	2	421	6.44 „	7. 1	7. 1	2 „	Davis' Strait . .	Greenland
11	3	422	6.45 „	7. 2	7. 2	3 „	Deal	England
11	4	423	6.46 „	7. 3	7. 3	4 „	Delagoa Bay . . .	Africa, E
11	5	424	6.47 „	7. 4	7. 4	5 „	Delaware, *Light*	United States, N. A.
11	6	425	6.48 „	7. 5	7. 5	6 „	Delgada, Pta., *Light*	St. Michael
11	7	426	6.49 „	7. 6	7. 6	7 „	Delgado, *Cape* . . .	Africa, E
11	8	427	6.50 „	7. 7	7. 7	8 „	Demarara .	South America
11	9	428	6.51 „	7. 8	7. 8	9 „	Deptford	England
11	10	429	6.52 „	7. 9	7. 9	10 „	Descada, *Cape*	Africa, S
11	13	430	6.53 „	7.10	7.10	11 „	Desirade, *Island* . .	West Indies
11	14	431	6 54 „	7.11	7.11	12 „	Diamond Point . . .	Sumatra
11	15	432	6.55 „	7.12	7.12	13 „	Diego Alvarez, *Island* .	Atlantic
11	16	433	6.56 „	7.13	7.13	14 „	Dieppe, *Light* . . .	France
11	17	434	6.57 „	7.14	7.14	15 „	Dingle Bay . . .	Ireland
11	18	435	6.58 „	7.15	7.15	16 „	Diu	Guzerat
11	19	436	6.59 „	7.16	7.16	17 „	Diursten, *Light* .	Gulf of Bothnia
11	20	437	7. 1 „	7.17	7.17	18 „	Doggersbank . .	North Sea
11	21	438	7. 2 „	7.18	7.18	19 „	Domel, *Island* .	Coast of Tanasserim
11	22	439	7. 3 „	7.19	7.19	20 „	Domenes, *Lights* . .	Russia
11	23	440	7. 4 „	7.20	7.20	21 „	Domingo, St. . .	Hayti
11	24	441	7. 5 „	7.21	7.21	22 „	Dominica, *Island* .	West Indies
11	25	442	7. 6 „	7.22	7.22	23 „	Dondree, *Cape* . . .	Ceylon
11	26	443	7. 7 „	7.23	7.23	24 „	Dory	New Guinea
11	27	444	7. 8 „	7.24	7.24	25 „	Dover Lights . . .	England
							Dover Straits, see Channel between Calais and Dover	
11	28	445	7. 9 „	7.25	7.25	26 „	Downs	England
11	29	446	7.10 „	7.26	7.26	27 „	Dragoe	Amack
11	30	447	7.11 „	7.27	7.27	28 „	Dram	Norway
							Drontheim, see Trondheim	
11	31	448	7.12 „	7.28	7.28	29 „	Dublin, *Light* . .	Ireland
11	32	449	7.13 „	7.29	7.29	30 „	Dudgeon, *Floating Light*	Hull, S E
11	33	450	7.14 „	7.30	7.30	——	Dulcigno	Albania
11	34	451	7.15 „	7.31	7.31	——	Dumfries . .	Scotland
11	35	452	7.16 „	7.32	7.32	——	Dunbar	Scotland
11	36	453	7.17 „	7.33	7.33	——	Duncansby-Head .	Scotland, N
11	37	454	7.18 „	7.34	7.34	——	Dundalk	Ireland
11	38	455	7.19 „	7.35	7.35	——	Dundee, *Lights* . . .	Scotland
11	39	456	7.20 „	7.36	7.36	——	Dungarvon . .	Ireland
11	40	457	7.21 „	7.37	7.37	——	Dungeness, *Light* .	England

COMPOUND TABLE.

Pair Signals. No.	No.	Numerals.	Time of the Day.	Latitude North.	Longitude East.	Days of the Month.	Nations, Countries, Seas, Seaports, Capes, &c.	
			H. M.	° ′	° ′			
12	1	458	7.22 AM.	7.38	7.38	1 Dec.	DUNKIRK, *Lights* . .	France
12	2	459	7.23 ,,	7.39	7.39	2 ,,	DUNNET-HEAD, *Light* .	Scotland
12	3	460	7.24 ,,	7.40	7.40	3 ,,	DUNNOSE	Wight
12	4	461	7.25 ,,	7.41	7.41	4 ,,	DURAZZO	Albania
12	5	462	7.26 ,,	7.42	7.42	5 ,,	EBELTOFTE	Jutland
12	6	463	7.27 ,,	7.43	7.43	6 ,,	ECKERNFOORDE . . .	Sleswick
12	7	464	7.28 ,,	7.44	7.44	7 ,,	EDDYSTONE, *Light* . .	England
12	8	465	7.29 ,,	7.45	7.45	8 ,,	EDDOE, *Light* . . .	North Sea
12	9	466	7.30 ,,	7.46	7.46	9 ,,	EGEDESMINDE . . .	Greenland
12	10	467	7.31 ,,	7.47	7.47	10 ,,	EGGERSUND	Norway
12	13	468	7.32 ,,	7.48	7.48	11 ,,	EGHOLM	Great Belt
12	14	469	7.33 ,,	7.49	7.49	12 ,,	EGMONT OP ZEE, *Lights* .	Holland
12	15	470	7.34 ,,	7.50	7.50	13 ,,	EGRIBOS . . .	Negropont
12	16	471	7.35 ,,	7.51	7.51	14 ,,	EGVAAG . . .	Norway
12	17	472	7.36 ,,	7.52	7.52	15 ,,	EIDER, *River* . .	Denmark
12	18	473	7.37 ,,	7.53	7.53	16 ,,	EIDER, *floating Light* .	Denmark
12	19	474	7.38 ,,	7.54	7.54	17 ,,	EKHOLM, *Light* .	Gulf of Finland
12	20	475	7.39 ,,	7.55	7.55	18 ,,	ELBA, *Island* . .	Mediterranean
12	21	476	7.40 ,,	7.56	7.56	19 ,,	ELBE, *River* .	Hanover & Holstein
12	22	477	7.41 ,,	7.57	7.57	20 ,,	ELBE, OUTER, *floating Light* .	do. do.
12	23	478	7.42 ,,	7.58	7.58	21 ,,	ELBE, INNER, *floating Light* .	do. do.
12	24	479	7.43 ,,	7.59	7.59	22 ,,	ELMINA . . .	Africa, W
							ELSINORE, see HELSINGOER	
12	25	480	7.44 ,,	8. 0	8. 0	23 ,,	EMBDEN . . .	Hanover
12	26	481	7.45 ,,	8. 1	8. 1	24 ,,	ENDELAVE, *Island* .	Cattegat
12	27	482	7.46 ,,	8. 2	8. 2	25 ,,	ENKHUYSEN, *Island* .	Iceland, E
12	28	483	7.47 ,,	8. 3	8. 3	26 ,,	ENSKAR, *Light* .	Gulf of Bothnia
12	29	484	7.48 ,,	8. 4	8. 4	27 ,,	ESPICHEL, *Cape, Light* .	Portugal
12	30	485	7.49 ,,	8. 5	8. 5	28 ,,	ESPRIT, ST., *Island* .	New Hebrides
12	31	486	7.50 ,,	8. 6	8. 6	29 ,,	ESQUIRQUES . .	Mediterranean
12	32	487	7.51 ,,	8. 7	8. 7	30 ,,	ESSEQUEBO .	South America
12	33	488	7.52 ,,	8. 8	8. 8	31 ,,	ETAPLES, *Lights* . .	France
12	34	489	7.53 ,,	8. 9	8. 9	——	EUROPA ROCKS .	Mozambique Channel
12	35	490	7.54 ,,	8.10	8.10	——	EUSTATIUS, ST., *Island* .	West Indies
12	36	491	7.55 ,,	8.11	8.11	——	EXMOUTH	England
12	37	492	7.56 ,,	8.12	8.12	——	FAABORG	Funen
12	38	493	7.57 ,,	8.13	8.13	——	FAIR-HILL . . .	Orkneys, NE
12	39	494	7.58 ,,	8.14	8.14	——	FAJUM . : . . .	Egypt
12	40	495	7.59 ,,	8.15	8.15	——	FAKKEBIERG, *Light*	Langeland, Denmark

COMPOUND TABLE.

Pair Signals. No.	Pair Signals. No.	Numerals.	Time of the Day.	Latitude North.	Longitude East.	Nations, Countries, Seas, Seaports, Capes, &c.	
			H. M.	° ′	°		
13	1	496	8. 1 A.M.	8.16	8.16	FALKENBERG	Sweden
13	2	497	8. 2 „	8.17	8.17	FALKLAND'S ISLANDS	America, S
13	3	498	8. 3 „	8.18	8.18	FALMOUTH	England
13	4	499	8. 4 „	8.19	8.19	FALSE BAY	Africa, S
13	5	500	8. 5 „	8.20	8.20	FALSE, *Cape*	Africa, S
13	6	501	8. 6 „	8.21	8.21	FALSTERBOE, *Light*	Sweden
13	7	502	8. 7 „	8.22	8.22	FALSTER, *Island*	Denmark
13	8	503	8. 8 „	8.23	8.23	FANNET, PT., *Light*	Ireland
13	9	504	8. 9 „	8.24	8.24	FANOE, *Island*	Denmark, W
13	10	505	8.10 „	8.25	8.25	FARSUND	Norway
13	11	506	8.11 „	8.26	8.26	FAREWELL, *Cape*	Greenland
13	12	507	8.12 „	8.27	8.27	FAROE	Gothland, NE
13	15	508	8.13 „	8.28	8.28	FAVIGNANA, *Island*	Sicily
13	16	509	8.14 „	8.29	8.29	FAXEFIORD	Iceland
13	17	510	8.15 „	8.30	8.30	FAYAL, *Island*	Azores
13	18	511	8.16 „	8.31	8.31	FEAR, *Cape, Light*	North America
13	19	512	8.17 „	8.32	8.32	FECAMP, *Light*	France
13	20	513	8.18 „	8.33	8.33	FELIX, ST., *Island*	Pacific Ocean
13	21	514	8.19 „	8.34	8.34	FELIX, ST., *Cape*	Sumatra
13	22	515	8.20 „	8.35	8.35	FEMERN, *Island, Light*	Holstein
13	23	516	8.21 „	8.36	8.36	FEMERNSUND	Femern, S
13	24	517	8.22 „	8.37	8.37	FEODOSIA (Kaffa) *Straits*	Taman & Krim
13	25	518	8.23 „	8.38	8.38	FERNANDO DEL PO	Africa, W
13	26	519	8.24 „	8.39	8.39	FERN, *Lights*	England
13	27	520	8.25 „	8.40	8.40	FERRAT, *Cape*	Africa, N
13	28	521	8.26 „	8.41	8.41	FERRO, *Island*	Canary Isles
13	29	522	8.27 „	8.42	8.42	FERROL	Spain
13	30	523	8.28 „	8.43	8.43	FIDSCHI ISLANDS	Australia
13	31	524	8.29 „	8.44	8.44	FIGUEIRA	Portugal
13	32	525	8.30 „	8.45	8.45	FILLEURI, *Island*	Sicily
13	33	526	8.31 „	8.46	8.46	FILSAND, *Lights*	Oesel
13	34	527	8.32 „	8.47	8.47	FINISTERRE, *Cape*	Spain
13	35	528	8.33 „	8.48	8.48	FINLAND	Russia
13	36	529	8.34 „	8.49	8.49	FINMARK	Norway, N
13	37	530	8.35 „	8.50	8.50	FINLAND, GULF OF	Russia
13	38	531	8.36 „	8.51	8.51	FIORENZO, ST.	Corsica
13	39	532	8.37 „	8.52	8.52	FLAMBOROUGH-HEAD, *Light*	England
13	40	533	8.38 „	8.53	8.53	FLATHOLM, *Light*	Bristol Channel

COMPOUND TABLE.

Pair Signals. No.	Pair Signals. No.	Numerals.	Time of the Day.	Latitude North.	Longitude East.	Nations, Countries, Seas, Seaports, Capes, &c.	
			H. M.	° ′	° ′		
14	1	534	8.39 AM.	8.54	8.54	FLEKKEFIORD . .	Norway
14	2	535	8.40 „	8.55	8.55	FLEKKEROE . . .	Norway
14	3	536	8.41 „	8.56	8.56	FLENSBURG . .	Sleswick
14	4	537	8.42 „	8.57	8.57	FLORES, *Island* . . .	Azores
14	5	538	8.43 „	8.58	8.58	FLORIDA . .	United States, N.A.
14	6	539	8.44 „	8.59	8.59	FOGO, *Island* . .	Cape Verd Islands
14	7	540	8.45 „	9. 0	9. 0	FOLKSTONE, *Light* . .	England
14	8	541	8.46 „	9. 1	9. 1	FORELAND, NORTH, *Light* .	England
14	9	542	8.47 „	9. 2	9. 2	FORELAND, SOUTH, *Lights* .	England
14	10	543	8.48 „	9. 3	9. 3	FORMENTERA, *Island* .	Mediterranean
14	11	544	8.49 „	9. 4	9. 4	FORMIGAS, *Rocks* . .	Azores, SE
14	12	545	8.50 „	9. 5	9. 5	FORMOSA, *Cape* . .	Africa, W
14	15	546	8.51 „	9. 6	9. 6	FORMOSA, OF TAIWAN, *Island* .	China
14	16	547	8.52 „	9. 7	9. 7	FORNŒS . . .	Cattegat
14	17	548	8.53 „	9. 8	9. 8	FORTAVENTURA, *Island* .	Canary Isles
14	18	549	8.54 „	9. 9	9. 9	FOUL, *Island* . .	Shetland Isles
14	19	550	8.55 „	9.10	9.10	FOUR, *Light* . . .	France
14	20	551	8.56 „	9.11	9.11	FRANCISCO, ST. . .	America, W
14	21	552	8.57 „	9.12	9.12	FRANÇOIS, *Cape* . .	Hayti, NE
14	22	553	8.58 „	9.13	9.13	FRECHEL, *Cape*, *Light* . .	France
14	23	554	8.59 „	9.14	9.14	FREDERICIA . . .	Jutland
14	24	555	9. 1 „	9.15	9.15	FREDERIKS-HAAB . .	Greenland
14	25	556	9. 2 „	9.16	9.16	FREDERIKSHALD .	Norway
14	26	557	9. 3 „	9.17	9.17	FREDERIKSHAMN . .	Finland
14	27	558	9. 4 „	9.18	9.18	FREDERIKS-HAVN, *Light* .	Jutland
14	28	559	9. 5 „	9.19	9.19	FREDERIKSTAD . .	Norway
14	29	560	9. 6 „	9.20	9.20	FREDERIKSTED . .	St. Croix
14	30	561	9. 7 „	9.21	9.21	FREDERIKSVŒRN . .	Norway
14	31	562	9. 8 „	9.22	9.22	FRIENDLY ISLANDS . .	Australia
14	32	563	9. 9 „	9.23	9.23	FRIESLAND . . .	Holland
14	33	564	9.10 „	9.24	9.24	FIRTH OF FORTH . .	Scotland
14	34	565	9.11 „	9.25	9.25	FROBISHER'S STRAITS .	Greenland
14	35	566	9.12 „	9.26	9.26	FROYEN, *Island* . .	Hitteren, N
14	36	567	9.13 „	9.27	9.27	FUGLEKLIPPEN (BIRD ROCK)	St. Thomas, SE
14	37	568	9.14 „	9.28	9.28	FULEHUK, *Light* .	Firth of Christiania
14	38	569	9.15 „	9.29	9.29	FUNCHALL . .	Madeira
14	39	570	9.16 „	9.30	9.30	FUNDY, BAY OF .	North America
14	40	571	9.17 „	9.31	9.31	FUNEN, *Island* . .	Denmark

COMPOUND TABLE.

Pair Signals. No.	Pair Signals. No.	Numerals.	Time of the Day.	Latitude North.	Longitude East	Nations, Countries, Seas, Seaports, Capes, &c.	
			H.M.	° ′	° ′		
15	1	572	9.18 AM.	9.32	9.32	FUNEN'S-HEAD	Funen, N.
15	2	573	9.19 ,,	9.33	9.33	FŒLLSK BAKE	Gulf of Bothnia
15	3	574	9.20 ,,	9.34	9.34	FŒRDER, *Light*	Firth of Christiania
15	4	575	9.21 ,,	9.35	9.35	FŒROE, *Islands*	Iceland, SE
15	5	576	9.22 ,,	9.36	9.36	FOEHR, *Island*	Denmark, W
15	6	577	9.23 ,,	9.37	9.37	GADDENS, *Light*	Gulf of Bothnia
15	7	578	9.24 ,,	9.38	9.38	GAETA	Naples
15	8	579	9.25 ,,	9.39	9.39	GAIR LOCH-BAY	Scotland, NW
15	9	580	9.26 ,,	9.40	9.40	GALLAN-HEAD	Lewis
15	10	581	9.27 ,,	9.41	9.41	GALLAPAGOS, *Islands*	Pacific Ocean
15	11	582	9.28 ,,	9.42	9.42	GALLIPOLI, *Light*	Turkey
15	12	583	9.29 ,,	9.43	9.43	GALLOPER, *Floating Light*	England
15	13	584	9.30 ,,	9.44	9.44	GALVESTON-BAY, *Light*	Gulf of Mexico
15	14	585	9.31 ,,	9.45	9.45	GALWAY	Ireland
15	17	586	9.32 ,,	9.46	9.46	GAMBIA, *River*	Africa, W
15	18	587	9.33 ,,	9.47	9.47	GATE, *Cape*	Spain
15	19	588	9.34 ,,	9.48	9.48	GEER, *Cape*	Africa, W
15	20	589	9.35 ,,	9.49	9.49	GEFLE	Sweden
15	21	590	9.36 ,,	9.50	9.50	GENOA	Italy
15	22	591	9.37 ,,	9.51	9.51	GEORGIA, WEST	America, NW
15	23	592	9.38 ,,	9.52	9.52	GEORGIA, SOUTH, *Island*	Atlantic
15	24	593	9.39 ,,	9.53	9.53	GEORGES, *Islands*	Pacific Ocean
15	25	594	9.40 ,,	9.54	9.54	GEORGESTOWN, *Light*	United States, N. A.
15	26	595	9.41 ,,	9.55	9.55	GEORGE'S CHANNEL, ST.	England and Ireland
15	27	596	9.42 ,,	9.56	9.56	GIBRALTAR	Spain
15	28	597	9.43 ,,	9.57	9.57	GIBRALTAR, STRAITS OF	Gibraltar, S
15	29	598	9.44 ,,	9.58	9.58	GIGLIO, *Island*	Italy
15	30	599	9.45 ,,	9.59	9.59	GIJON	Spain
15	31	600	9.46 ,,	10. 0	10. 0	GIRGENTI, *Light*	Sicily
15	32	601	9.47 ,,	10. 1	10. 1	GIRONDE, *River*, *Lights*	France
15	33	602	9.48 ,,	10. 2	10. 2	GIEDSER, *Light*	Falster
15	34	603	9.49 ,,	10. 3	10. 3	GLASGOW	Scotland
15	35	604	9.50 ,,	10. 4	10. 4	GLUCKSTADT	Holstein
15	36	605	9.51 ,,	10. 5	10. 5	GOA	Hindostan
15	37	606	9.52 ,,	10. 6	10. 6	GODTHAAB	Greenland
15	38	607	9.53 ,,	10. 7	10. 7	GOEREE, *Light*	Holland
15	39	608	9.54 ,,	10. 8	10. 8	GOMERA, *Island*	Canary Isles
15	40	609	9.55 ,,	10. 9	10. 9	GOODWIN SAND, *Floating Lights*	Channel

COMPOUND TABLE.

Pair Signals. No.	No.	Numerals.	Time of the Day. H. M.	Latitude North. ° ′	Longitude East. ° ′	Nations, Countries, Seas, Seaports, Capes, &c.	
16	1	610	9.56 A.M.	10.10	10.10	Goodwin Sand, S. Sand Head, *fl. Light*	England
16	2	611	9.57 ,,	10.11	10.11	Gotha Canal	Sweden
16	3	612	9.58 ,,	10.12	10.12	Gothenborg	Sweden
16	4	613	9.59 ,,	10.13	10.13	Gottska-Sandoe	Gothland, N
16	5	614	10. 1 ,,	10.14	10.14	Gough's Islands	Atlantic
16	6	615	10. 2 ,,	10.15	10.15	Gozzo, *Island*	Malta, NW
16	7	616	10. 3 ,,	10.16	10.16	Gracias a dios	Honduras
16	8	617	10. 4 ,,	10.17	10.17	Graciosa, *Island*	Azores
16	9	618	10. 5 ,,	10.18	10.18	Grenadines, *Islands*	West Indies
16	10	619	10. 6 ,,	10.19	10.19	Grande, Rio	South America
16	11	620	10. 7 ,,	10.20	10.20	Grangemouth	Scotland
16	12	621	10. 8 ,,	10.21	10.21	Granville, *Rocks, Lights*	France
16	13	622	10. 9 ,,	10.22	10.22	Grave, Point de, *Light*	France
16	14	623	10.10 ,,	10.23	10.23	Gravelines, *Light*	France
16	17	624	10.11 ,,	10.24	10.24	Gravesend	England
16	18	625	10.12 ,,	10.25	10.25	Greenock	Scotland
16	19	626	10.13 ,,	10.26	10.26	Greenwich, *Observatory*	England
16	20	627	10.14 ,,	10.27	10.27	Greifswaldoe, *Light*	Rugen, SE
16	21	628	10.15 ,,	10.28	10.28	Grenaae	Jutland
16	22	629	10.16 ,,	10.29	10.29	Grenada, *Island*	West Indies
16	23	630	10.17 ,,	10.30	10.30	Grimsby	England
16	24	631	10.18 ,,	10.31	10.31	Grimstad	Norway
16	25	632	10.19 ,,	10.32	10.32	Grisnez, *Cape, Light*	France
16	26	633	10.20 ,,	10.33	10.33	Groeningen	Holland
16	27	634	10.21 ,,	10.34	10.34	Groix, Isle de, *Light*	France, W
16	28	635	10.22 ,,	10.35	10.35	Greenland	America, N
16	29	636	10.23 ,,	10.36	10.36	Groenskœr, *Light*	Stockholm, E
16	30	637	10.24 ,,	10.37	10.37	Groensund, *Passage*	Falster & Moen
16	31	638	10.25 ,,	10.38	10.38	Guadaloupe, *Island*	West Indies
16	32	639	10.26 ,,	10.39	10.39	Guam, *Island*	Marianne Islands
16	33	640	10.27 ,,	10.40	10.40	Guardafui, *Cape*	Africa, E
16	34	641	10.28 ,,	10.41	10.41	Guatimala	America, M
16	35	642	10.29 ,,	10.42	10.42	Guayaquil	America, W
16	36	643	10.30 ,,	10.43	10.43	Guernsey, *Island*	Channel
16	37	644	10.31 ,,	10.44	10.44	Guia, Na. Sa. de, *Light*	Portugal
16	38	645	10.32 ,,	10.45	10.45	Guinea	Africa, W
16	39	646	10.33 ,,	10.46	10.46	Guldborgsund, *Passage*	Falster & Laaland
16	40	647	10.34 ,,	10.47	10.47	Gold Coast	Africa, W

COMPOUND TABLE.

Pair Signals. No	Pair Signals. No.	Numerals.	Time of the Day.	Latitude North.	Longitude East.	Nations, Countries, Seas, Seaports, Capes, &c.	
			H. M.	° ′	° ′		
17	1	648	10.35 AM.	10.48	10.48	Gull, *floating Light*	Channel
17	2	649	10.36 ,,	10.49	10.49	Gothland, *Island*	Baltic
17	3	650	10.37 ,,	10.50	10.50	Hague	Holland
17	4	651	10.38 ,,	10.51	10.51	Hacha, Rio de la	S. America
17	5	652	10.39 ,,	10.52	10.52	Hainan, *Island*	China
17	6	653	10.40 ,,	10.53	10.53	Halifax	Nova Scotia
17	7	654	10.41 ,,	10.54	10.54	Hallands-Vœderoe	Sweden
17	8	655	10.42 ,,	10.55	10.55	Halmstad	Sweden
17	9	656	10.43 ,,	10.56	10.56	Halskov Reef, *Light*	Sealand, W
17	10	657	10.44 ,,	10.57	10.57	Hammersfest	Finmark
17	11	658	10.45 ,,	10.58	10.58	Hammerhuus, *Light*	Bornholm
17	12	659	10.46 ,,	10.59	10.59	Hanstholmen	Jutland, NW
17	13	660	10.47 ,,	11. 0	11. 0	Hanoe	Carlshamn, S
17	14	661	10.48 ,,	11. 1	11. 1	Harlingen	Holland
17	15	662	10.49 ,,	11. 2	11. 2	Harwich, *Lights*	England
17	16	663	10.50 ,,	11. 3	11. 3	Hasborough, *Lights*	England, E
17	19	664	10.51 ,,	11. 4	11. 4	Hasborough Gat, *floating Lights*	England, E
17	20	665	10.52 ,,	11. 5	11. 5	Hasborough Sand, *floating Light*	England, E
17	21	666	10.53 ,,	11. 6	11. 6	Hasteen's Ground	Cattegat
17	22	667	10.54 ,,	11. 7	11. 7	Hastings, *Lights*	England
17	23	668	10.55 ,,	11. 8	11. 8	Hatteras, *Cape, Light*	U. States, N.A.
17	24	669	10.56 ,,	11. 9	11. 9	Havannah	Cuba
17	25	670	10.57 ,,	11.10	11.10	Havre de Grace, *Light*	France
17	26	671	10.58 ,,	11.11	11.11	Hayti, (St. Domingo), *Island*	West Indies
17	27	672	10.59 ,,	11.12	11.12	Hayti, *Cape*	Hayti
17	28	673	11. 1 ,,	11.13	11.13	Hebrides	Scotland, W
17	29	674	11. 2 ,,	11.14	11.14	Hedio, *Island, Light*	France, W
17	30	675	11. 3 ,,	11.15	11.15	Heel, *Light*	Gulf of Dantzic
17	31	676	11. 4 ,,	11.16	11.16	Heiligenhafen	Holstein
17	32	677	11. 5 ,,	11.17	11.17	Helder, *Light*	Holland
17	33	678	11. 6 ,,	11.18	11.18	Helena, St., *Island*	Atlantic
17	34	679	11. 7 ,,	11.19	11.19	Helen's, St.	Wight
17	35	680	11. 8 ,,	11.20	11.20	Heligoland, *Island, Light*	North Sea
17	36	681	11. 9 ,,	11.21	11.21	Helsingborg	Sweden
17	37	682	11.10 ,,	11.22	11.22	Helsingfors	Finland
17	38	683	11.11 ,,	11.23	11.23	Helsingoer, *Cronborg Light*	Sealand
17	39	684	11.12 ,,	11.24	11.24	Helvoetsluys	Holland
17	40	685	11.13 ,,	11.25	11.25	Henlopen, *Cape*	U. States, N.A.

COMPOUND TABLE.

Pair Signals No.	Pair Signals No.	Numerals.	Time of the Day.	Latitude North.	Longitude East.	Nations, Countries, Seas, Seaports, Capes, &c.	
			H. M.	° ′	° ′		
18	1	686	11.14 A.M.	11.26	11.26	Henry, *Cape*, *Light* .	U. States, N. A.
18	2	687	11.15 ,,	11.27	11.27	Hernhut, New . .	Greenland
18	3	688	11.16 ,,	11.28	11.28	Hernoesand . . .	Sweden
18	4	689	11.17 ,,	11.29	11.29	Hesseloe, *Island* . .	Cattegat
18	5	690	11.18 ,,	11.30	11.30	Heve, Cape la, *Lights* . .	France
18	6	691	11.19 ,,	11.31	11.31	Hielm, *Island* . .	Cattegat
18	7	692	11.20 ,,	11.32	11.32	Hind, *Island* . .	Northland
18	8	693	11.21 ,,	11.33	11.33	Hirtshals . .	Jutland, N W
18	9	694	11.22 ,,	11.34	11.34	Hirtsholmene . . .	Cattegat
18	10	695	11.23 ,,	11.35	11.35	Hitteren, *Island* . .	Norway, W
18	11	696	11.24 ,,	11.36	11.36	Hoang-Hay . . .	China
18	12	697	11.25 ,,	11.37	11.37	Hoborg . . .	Gothland
18	13	698	11.26 ,,	11.38	11.38	Hogland, *Island*, *Light*	Gulf of Finland
18	14	699	11.27 ,,	11.39	11.39	Hogue, Cape la . .	France
18	15	700	11.28 ,,	11.40	11.40	Hollander-deep . .	The Sound
18	16	701	11.29 ,,	11.41	11.41	Holmestrand . . .	Norway
18	19	702	11.30 ,,	11.42	11.42	Holstein . . .	Denmark
18	20	703	11.31 ,,	11.43	11.43	Holyhead, *Light* .	England, W
18	21	704	11.32 ,,	11.44	11.44	Honduras . . .	America, M
18	22	705	11.33 ,,	11.45	11.45	Honfleur, *Lights* . .	France
18	23	706	11.34 ,,	11.46	11.46	Horn, *Cape* . .	America, S
18	24	707	11.35 ,,	11.47	11.47	Horn's Reef . . .	Jutland
18	25	708	11.36 ,,	11.48	11.48	Howe's Islands . . .	Australia
18	26	709	11.37 ,,	11.49	11.49	Hudickswall . . .	Sweden
18	27	710	11.38 ,,	11.50	11.50	Hudson's Straits .	North America
18	28	711	11.39 ,,	11.51	11.51	Hull	England
18	29	712	11.40 ,,	11.52	11.52	Hundholmen . .	Northland
18	30	713	11.41 ,,	11.53	11.53	Hunstanton, *Light* . .	England
18	31	714	11.42 ,,	11.54	11.54	Hurst, *Lights* . .	England
18	32	715	11.43 ,,	11.55	11.55	Husum . . .	Sleswick
18	33	716	11.44 ,,	11.56	11.56	Hvaloerne . .	Skagerak
18	34	717	11.45 ,,	11.57	11.57	Hven, *Island* . .	The Sound
18	35	718	11.46 ,,	11.58	11.58	Hviddingsoe, *Light* . .	Karmoe
18	36	719	11.47 ,,	11.59	11.59	Hvide Hav, *or* White Sea	Russia, N
18	37	720	11.48 ,,	——	——	Hydra, *Island* . .	Morea, E
18	38	721	11.49 ,,	——	——	Hyeres, de, *Islands* . .	France
18	39	722	11.50 ,,	——	——	Hœfringe-Bake	Bay of Norrkoeping
18	40	723	11.51 ,,	——	——	Hœradskœr-Bake	ditto ditto

COMPOUND TABLE.

Pair Signals. No.	Pair Signals. No.	Numerals.	Time of the Day. H.M.	Latitude South. ° ′	Longitude West. ° ′	Nations, Countries, Seas, Seaports, Capes, &c.	
19	1	724	11.52 AM.	0. 1	0. 1	Hoegsten's Bake	Sea of Aland
19	2	725	11.53 ,,	0. 2	0. 2	Hoievarde, *Light*	Norway
19	3	726	11.54 ,,	0. 3	0. 3	Jackson, Port	New South Wales
19	4	727	11.55 ,,	0. 4	0. 4	Jago, St., *Island*	Cape Verd Isles
19	5	728	11.56 ,,	0. 5	0. 5	Jago, St.	Cuba
19	6	729	11.57 ,,	0. 6	0. 6	Jamaica, *Island*	West Indies
19	7	730	11.58 ,,	0. 7	0. 7	James's, St., *Cape*	Tsiompa
19	8	731	11.59 ,,	0. 8	0. 8	Jan, St., *or* St. John's, *Island*	West Indies
19	9	732	12. 1 PM.	0. 9	0. 9	Janeiro, Rio de	Brazil
19	10	733	12. 2 ,,	0.10	0.10	Japan Sea	Japan
19	11	734	12. 3 ,,	0.11	0.11	Jaquemel	Hayti
19	12	735	12. 4 ,,	0.12	0.12	Java, *Island*	Indian Ocean
19	13	736	12. 5 ,,	0.13	0.13	I-Colm-kill, *Island*	Hebrides
19	14	737	12. 6 ,,	0.14	0.14	Jean d'Acre, St.	Syria
19	15	738	12. 7 ,,	0.15	0.15	Jean de Luz, St. *Light*	France
19	16	739	12. 8 ,,	0.16	0.16	Jedderen	Norway
19	17	740	12. 9 ,,	0.17	0.17	Jersey, *Island*	Channel
19	18	741	12.10 ,,	0.18	0.18	Ilfracombe, *Light*	England
19	21	742	12.11 ,,	0.19	0.19	Imaglin	Behring's Straits
19	22	743	12.12 ,,	0.20	0.20	Indies, East	Asia
19	23	744	12.13 ,,	0.21	0.21	Indies, West	America
19	24	745	12.14 ,,	0.22	0.22	Ingen	Finmark
19	25	746	12.15 ,,	0.23	0.23	Insterholl, *Island, Light*	Ireland, N
19	26	747	12.16 ,,	0.24	0.24	Inverness	Scotland
19	27	748	12.17 ,,	0.25	0.25	John's, St.	Newfoundland
19	28	749	12.18 ,,	0.26	0.26	John de Nova, *Island*	Indian Ocean
19	29	750	12.19 ,,	0.27	0.27	Ipsara, *Island*	Archipelago
19	30	751	12.20 ,,	0.28	0.28	Ireland, *Island*	Great Britain
19	31	752	12.21 ,,	0.29	0.29	Irkutsk	Russia in Asia
19	32	753	12.22 ,,	0.30	0.30	Ischia, *Island*	Italy
19	33	754	12.23 ,,	0.31	0.31	Isefiord	Sealand
19	34	755	12.24 ,,	0.32	0.32	Icy Sea, Arctic	North Pole
19	35	756	12.25 ,,	0.33	0.33	Icy Sea, Antarctic	South Pole
19	36	757	12.26 ,,	0.34	0.34	Iceland, *Island*	Greenland, E
19	37	758	12.27 ,,	0.35	0.35	Isle de Bas	France, W
19	38	759	12.28 ,,	0.36	0.36	Isolette, *Cape*	Arabia
19	39	760	12.29 ,,	0.37	0.37	Juan Fernandez, *Light*	Pacific Ocean
19	40	761	12.30 ,,	0.38	0.38	Juan, St.	Porto-Rico

COMPOUND TABLE.

Pair Signals. No.	No.	Numerals.	Time of the Day.	Latitude South.	Longitude West.	Nations, Countries, Seas, Seaports, Capes, &c.
			H. M.	° ′	° ′	
20	1	762	12.31 P.M.	0.39	0.39	Julianehaab . . . . . Greenland
20	2	763	12.32 ,,	0.40	0.40	Ivitza, *Island* . . . Mediterranean
20	3	764	12.33 ,,	0.41	0.41	Jutland . . . . . . Denmark
20	4	765	12.34 ,,	0.42	0.42	Kaffa . . . . . Russia
20	5	766	12.35 ,,	0.43	0.43	Kalleboerne, *Passage* . Amack & Sealand
20	6	767	12.36 ,,	0.44	0.44	Kallundborg . . . . . Sealand
20	7	768	12.37 ,,	0.45	0.45	Kalmar . . . . Sweden
20	8	769	12.38 ,,	0.46	0.46	Kamtschatka . . . . . . Asia
20	9	770	12.39 ,,	0.47	0.47	Karskoi Sea . . . . . Siberia
20	10	771	12.40 ,,	0.48	0.48	Karmoe, *Island* . . . Norway, W
20	11	772	12.41 ,,	0.49	0.49	Kattegat . . . Denmark & Sweden
20	12	773	12.42 ,,	0.50	0.50	Kennebeck . . . United States, N.A.
20	13	774	12.43 ,,	0.51	0.51	Kerguelens Island, *or* Desolation, Indian Ocean
20	14	775	12.44 ,,	0.52	0.52	Key-West . . . United States, N.A.
20	15	776	12.45 ,,	0.53	0.53	Kiel . . . . . . Holstein
20	16	777	12.46 ,,	0.54	0.54	Kilda, St., *Island* . . Hebrides, W
20	17	778	12.47 ,,	0.55	0.55	Kingston . . . . . Jamaica
20	18	779	12.48 ,,	0.56	0.56	Kinnaird's Head, *Light* . . Scotland
20	21	780	12.49 ,,	0.57	0.57	Kinsale, *Light* . . . Ireland
20	22	781	12.50 ,,	0.58	0.58	Kinsale, Old Head of, *Light* . Ireland
20	23	782	12.51 ,,	0.59	0.59	Kiobenhavn . . . Denmark
20	24	783	12.52 ,,	1. 0	1. 0	Kioge Bay . . . . Denmark
20	25	784	12.53 ,,	1. 1	1. 1	Kniphausen . . . Oldenborg
20	26	785	12.54 ,,	1. 2	1. 2	Knud's Head, *Light* . . . Funen
20	27	786	12.55 ,,	1. 3	1. 3	Kokskœr, *Light* . . Gulf of Finland
20	28	787	12.56 ,,	1. 4	1. 4	Kola . . . . . Russia, N
20	29	788	12.57 ,,	1. 5	1. 5	Kolding . . . . . Jutland
20	30	789	12.58 ,,	1. 6	1. 6	Kongedybet (Kings deep) . The Sound
20	31	790	12.59 ,,	1. 7	1. 7	Kopparstenarne . . Gothland, N
20	32	791	1. 1 ,,	1. 8	1. 8	Korsoe, *Light* . . Stockholm, E
20	33	792	1. 2 ,,	1. 9	1. 9	Kotelnoi . . Arctic-Icy-Ocean
20	34	793	1. 3 ,,	1.10	1.10	Kotzebue Sound . . America, NW
20	35	794	1. 4 ,,	1.11	1.11	Krageroe . . . Norway
20	36	795	1. 5 ,,	1.12	1.12	Kruzenstern, *Cape* . . America, NW
20	37	796	1. 6 ,,	1.13	1.13	Kullen, *Light* . . . Sweden
20	38	797	1. 7 ,,	1.14	1.14	Kurili Islands . . Russia in Asia
20	39	798	1. 8 ,,	1.15	1.15	Kyholm (*Quarantine*), *Light* . Cattegat
20	40	799	1. 9 ,,	1.16	1.16	Kykduin, *Light* . . . Holland

COMPOUND TABLE.

Pair Signals. NO	NO	Numerals.	Time of the Day.	Latitude South.	Longitude West.	Nations, Countries, Seas, Seaports, Capes, &c.	
			H.M.	° ′	° ′		
21	1	800	1.10 P.M.	1.17	1.17	Kænguru, *Island*	New Holland, S
21	2	801	1.11 „	1.18	1.18	Koenigsberg	Prussia
21	3	802	1.12 „	1.19	1.19	Laaland, *Island*	Denmark
21	4	803	1.13 „	1.20	1.20	Labrador	North America
21	5	804	1.14 „	1.21	1.21	Laccadive Islands	Indian Ocean
21	6	805	1.15 „	1.22	1.22	Lagosta, *Island*	Adriatic
21	7	806	1.16 „	1.23	1.23	Lagskœr-Bake	Sea of Aland
21	8	807	1.17 „	1.24	1.24	Laguaira	South America
21	9	808	1.18 „	1.25	1.25	Lambness	Shetland Isles, N
21	10	809	1.19 „	1.26	1.26	Lampedusa, *Island*	Mediterranean
21	11	810	1.20 „	1.27	1.27	Lancaster, *Light*	England
21	12	811	1.21 „	1.28	1.28	Land's end, *Light*	England
21	13	812	1.22 „	1.29	1.29	Landscrona	Sweden
21	14	813	1.23 „	1.30	1.30	Landsort, *Light*	Bay of Norrkoeping
21	15	814	1.24 „	1.31	1.31	Langeland, *Island*	Denmark
21	16	815	1.25 „	1.32	1.32	Langesund	Norway
21	17	816	1.26 „	1.33	1.33	Lanzarote, *Island*	Canary Isles
21	18	817	1.27 „	1.34	1.34	Latham's Ground	Africa, E
21	19	818	1.28 „	1.35	1.35	Lavensari, *Island*	Gulf of Finland
21	20	819	1.29 „	1.36	1.36	Lawrence, St., *River*	North America
21	23	820	1.30 „	1.37	1.37	Leith, *Light*	Scotland
21	24	821	1.31 „	1.38	1.38	Lerwick	Shetland Isles
21	25	822	1.32 „	1.39	1.39	Leuca, Cape S. M^a de	Italy
21	26	823	1.33 „	1.40	1.40	Lewis, *Island*	Hebrides
21	27	824	1.34 „	1.41	1.41	Libau	Russia
21	28	825	1.35 „	1.42	1.42	Liberia	Guinea
21	29	826	1.36 „	1.43	1.43	Lichtenfeld	Greenland
21	30	827	1.37 „	1.44	1.44	Lillesand	Norway
21	31	828	1.38 „	1.45	1.45	Lima	Peru
21	32	829	1.39 „	1.46	1.46	Limerick	Ireland
21	33	830	1.40 „	1.47	1.47	Lindersnœs, *Light*	Norway, S
21	34	831	1.41 „	1.48	1.48	Lipari Islands	Sicily
21	35	832	1.42 „	1.49	1.49	Lisburne, *Cape*	America, NW
21	36	833	1.43 „	1.50	1.50	Lissa	Dalmatia
21	37	834	1.44 „	1.51	1.51	Lisbon	Portugal
21	38	835	1.45 „	1.52	1.52	Liverpool	England
21	39	836	1.46 „	1.53	1.53	Leghorn, *Light*	Tuscany
21	40	837	1.47 „	1.54	1.54	Lizard, *Cape, Lights*	England

COMPOUND TABLE.

Pair Signals. No.	Pair Signals. No.	Numerals.	Time of the Day.	Latitude South.	Longitude West.	Nations, Countries, Seas, Seaports, Capes, &c.	
			H. M.	° ′	° ′		
22	1	838	1.48 PM.	1.55	1.55	Loango	Africa, W
22	2	839	1.49 ,,	1.56	1.56	London	England
22	3	840	1.50 ,,	1.57	1.57	Long-Island	U. States, N.A.
22	4	841	1.51 ,,	1.58	1.58	Look-out, *Cape, Light*	U. States, N.A.
22	5	842	1.52 ,,	1.59	1.59	Loop-Head, *Light*	Ireland
22	6	843	1.53 ,,	2. 0	2. 0	Lopatka, *Cape*	Kamschatka
22	7	844	1.54 ,,	2. 1	2. 1	L'Orient	France
22	8	845	1.55 ,,	2. 2	2. 2	Louisbourg	Nova Scotia
22	9	846	1.56 ,,	2. 3	2. 3	Louisiade, *Islands*	Australia
22	10	847	1.57 ,,	2. 4	2. 4	Lovisa	Finland
22	11	848	1.58 ,,	2. 5	2. 5	Lowestoffe, *Lights*	England
22	12	849	1.59 ,,	2. 6	2. 6	Lucar de Baromeda, St.	Spain
22	13	850	2. 1 ,,	2. 7	2. 7	Lucas, St., *Cape*	America, W
22	14	851	2. 2 ,,	2. 8	2. 8	Lucia, St., *Island*	West Indies
22	15	852	2. 3 ,,	2. 9	2. 9	Luz, Nª Sª de la, *Light*	Portugal
22	16	853	2. 4 ,,	2.10	2.10	Lundy, *Island, Light*	Bristol Channel
22	17	854	2. 5 ,,	2.11	2.11	Lymington	England
22	18	855	2. 6 ,,	2.12	2.12	Luneburg	Hanover
22	19	856	2. 7 ,,	2.13	2.13	Lynn-Regis, *Light*	England
22	20	857	2. 8 ,,	2.14	2.14	Lynn Well, *floating Light*	England
22	23	858	2. 9 ,,	2.15	2.15	Lysegrund	Cattegat
22	24	859	2.10 ,,	2.16	2.16	Læsoe, *Island*	Cattegat
22	25	860	2.11 ,,	2.17	2.17	Læsoe, North-west reef of	Cattegat
22	26	861	2.12 ,,	2.18	2.18	Macao	China
22	27	862	2.13 ,,	2.19	2.19	Macassar	Celebes
22	28	863	2.14 ,,	2.20	2.20	Madagascar, *Island*	Africa, E
22	29	864	2.15 ,,	2.21	2.21	Madeira, *Island*	Atlantic
22	30	865	2.16 ,,	2.22	2.22	Madras, *Light*	Coast of Coromandel
22	31	866	2.17 ,,	2.23	2.23	Madura, *Island*	Java, NE
22	32	867	2.18 ,,	2.24	2.24	Magadoxa	Africa, E
22	33	868	2.19 ,,	2.25	2.25	Magellan's Straits	America, S
22	34	869	2.20 ,,	2.26	2.26	Mageroe	Finmark
22	35	870	2.21 ,,	2.27	2.27	Magindanao, *or* Mindanao	Philippine Isles
22	36	871	2.22 ,,	2.28	2.28	Mahe, *or* Sechelles Islands	Madagascar, NE
22	37	872	2.23 ,,	2.29	2.29	Mahon, Port	Minorca
22	38	873	2.24 ,,	2.30	2.30	Malacca	East Indies
22	39	874	2.25 ,,	2.31	2.31	Malacca, Straits of	Malacca
22	40	875	2.26 ,,	2.32	2.32	Maladives, *Islands*	Indian Ocean

COMPOUND TABLE.

Pair Signals. No.	Pair Signals. No.	Numerals.	Time of the Day. H. M.	Latitude South. ° ′	Longitude West. ° ′	Nations, Countries, Seas, Seaports, Capes, &c.	
23	1	876	2.27 PM.	2.33	2.33	Malaga, *Light*	Spain
23	2	877	2.28 „	2.34	2.34	Malay	East Indies
23	3	878	2.29 „	2.35	2.35	Majorca, *Island*	Mediterranean
23	4	879	2.30 „	2.36	2.36	Malmoe	Sweden
23	5	880	2.31 „	2.37	2.37	Malo, St.	France
23	6	881	2.32 „	2.38	2.38	Malta, *Island*	Mediterranean
23	7	882	2.33 „	2.39	2.39	Maloe-Bake	Gulf of Bothnia
23	8	883	2.34 „	2.40	2.40	Mandal	Norway
23	9	884	2.35 „	2.41	2.41	Mangalore	Malabar Coast
23	10	885	2.36 „	2.42	2.42	Manilla	Philippine Isles
23	11	886	2.37 „	2.43	2.43	Man, Isle of	Irish Sea
23	12	887	2.38 „	2.44	2.44	Mantavai, *or* Si Biru, *Island*	Sumatra, W
23	13	888	2.39 „	2.45	2.45	Maracaybo	South America
23	14	889	2.40 „	2.46	2.46	Maranon, Rio	South America
23	15	890	2.41 „	2.47	2.47	Marble Head	U. States, N.A.
23	16	891	2.42 „	2.48	2.48	Margarita, *Island*	South America
23	17	892	2.43 „	2.49	2.49	Margate, *Light*	England
23	18	893	2.44 „	2.50	2.50	Marianne Islands, *or* Ladrones	Australia
23	19	894	2.45 „	2.51	2.51	Maria, St., *Island*	Azores
23	20	895	2.46 „	2.52	2.52	Maria, St. *Cape*	Madagascar
23	21	896	2.47 „	2.53	2.53	Mariegalante, *Island*	West Indies
23	22	897	2.48 „	2.54	2.54	Markoe, *Light*	Norway
23	25	898	2.49 „	2.55	2.55	Marlborough Fort	Sumatra
23	26	899	2.50 „	2.56	2.56	Marmora, Sea of	Dardanelles, E
23	27	900	2.51 „	2.57	2.57	Marquesas, *Islands*	Australia
23	28	901	2.52 „	2.58	2.58	Marseille	France
23	29	902	2.53 „	2.59	2.59	Marstrand, *Light*	Sweden
23	30	903	2.54 „	3. 0	3. 0	Martaban	East Indies
23	31	904	2.55 „	3. 1	3. 1	Martha, St.	South America
23	32	905	2.56 „	3. 2	3. 2	Martha's Vineyard, *Island*	U. States, N.A.
23	33	906	2.57 „	3. 3	3. 3	Martin, St., *Island*	West Indies
23	34	907	2.58 „	3. 4	3. 4	Martinique, *Island*	West Indies
23	35	908	2.59 „	3. 5	3. 5	Massa	Italy
23	36	909	3. 1 „	3. 6	3. 6	Masulipatam	Golconda
23	37	910	3. 2 „	3. 7	3. 7	Matanzas	Cuba
23	38	911	3. 3 „	3. 8	3. 8	Matapan, *Cape*	Morea
23	39	912	3. 4 „	3. 9	3. 9	Mathieu, St., *Light*	France
23	40	913	3. 5 „	3.10	3.10	Maura, St., *Island*	Ionian Sea

COMPOUND TABLE.

Pair Signals No.	Pair Signals No.	Numerals.	Time of the Day. H. M.	Latitude South. ° ′	Longitude West. ° ′	Nations, Countries, Seas, Seaports, Capes, &c.	
24	1	914	3. 6 PM.	3.11	3.11	Mauritius, *Island* . .	Indian Ocean
24	2	915	3. 7 ,,	3.12	3.12	May, Isle of, *Light* .	Edinburgh, NE
24	3	916	3. 8 ,,	3.13	3.13	Mayo, *Island* . .	Cape Verd Isles
24	4	917	3. 9 ,,	3.14	3.14	Mayotta, *Island* .	Mosambique Channel
24	5	918	3.10 ,,	3.15	3.15	Mazzara . . .	Sicily
24	6	919	3.11 ,,	3.16	3.16	Mediterranean	
24	7	920	3.12 ,,	3.17	3.17	Melazzo, *Light* . . .	Sicily
24	8	921	3.13 ,,	3.18	3.18	Meldorf . . .	Holstein
24	9	922	3.14 ,,	3.19	3.19	Memel, *Light* . . .	Prussia
24	10	923	3.15 ,,	3.20	3.20	Mendocino, *Cape* . .	America, W
24	11	924	3.16 ,,	3.21	3.21	Messina, *Light* . . .	Sicily
24	12	925	3.17 ,,	3.22	3.22	Mesurate, *Cape* . . .	Africa, N
24	13	926	3.18 ,,	3.23	3.23	Metelino, *Island* . .	Archipelago
24	14	927	3.19 ,,	3.24	3.24	Mexico, Gulf of . .	America, M
24	15	928	3.20 ,,	3.25	3.25	Middleground, Little . .	Cattegat
24	16	929	3.21 ,,	3.26	3.26	Middleground . .	The Sound
24	17	930	3.22 ,,	3.27	3.27	Milford Haven, *St. Anne's Lights* .	England
24	18	931	3.23 ,,	3.28	3.28	Michael's, St., *Island* . .	Azores
24	19	932	3.24 ,,	3.29	3.29	Milo, *Island* . . .	Archipelago
24	20	933	3.25 ,,	3.30	3.30	Minorca, *Island* . . .	Mediterranean
24	21	934	3.26 ,,	3.31	3.31	Meinguiers, Les . .	France, W
24	22	935	3.27 ,,	3.32	3.32	Missisippi, *Light* . .	U. States, N. A.
24	25	936	3.28 ,,	3.33	3.33	Mobile, *Light* . .	U. States, N. A.
24	26	937	3.29 ,,	3.34	3.34	Modon	Morea
24	27	938	3.30 ,,	3.35	3.35	Mogadore	Morocco
24	28	939	3.31 ,,	3.36	3.36	Mokka	Arabia
24	29	940	3.32 ,,	3.37	3.37	Molucca Islands . .	East Indies
24	30	941	3.33 ,,	3.38	3.38	Mona Strait . . .	Porto-Rico, W
24	31	942	3.34 ,,	3.39	3.39	Monfia, *Island* . . .	Africa, E
24	32	943	3.35 ,,	3.40	3.40	Mont Christo, *Island* . .	Italy, W
24	33	944	3.36 ,,	3.41	3.41	Montego Bay . . .	Jamaica
24	34	945	3.37 ,,	3.42	3.42	Monteverdos Islands . .	Australia
24	35	946	3.38 ,,	3.43	3.43	Montevideo, *Light* . .	Rio de la Plata
24	36	947	3.39 ,,	3.44	3.44	Montpellier . . .	France
24	37	948	3.40 ,,	3.45	3.45	Montserrat, *Island* . .	West Indies
24	38	949	3.41 ,,	3.46	3.46	Montrose, *Lights* . . .	Scotland
24	39	950	3.42 ,,	3.47	3.47	Montuck, *Light* . .	U. States, N. A.
24	40	951	3.43 ,,	3.48	3.48	Morant, *Cape* . . .	Jamaica

COMPOUND TABLE.

Pair Signals. No.	No.	Numerals.	Time of the Day.	Latitude South.	Longitude West.	Nations, Countries, Seas, Seaports, Capes, &c.	
			H. M.	° ′	° ′		
25	1	952	3.44 PM.	3.49	3.49	Morea	Greece
25	2	953	3.45 ,,	3.50	3.50	Morebat, *Cape*	Arabia
25	3	954	3.46 ,,	3.51	3.51	Morlaix, *Light*	France
25	4	955	3.47 ,,	3.52	3.52	Moskoe (*Malstrom*)	Finmark
25	5	956	3.48 ,,	3.53	3.53	Mozambique	Africa, E
25	6	957	3.49 ,,	3.54	3.54	Mugford, *Cape*	North America
25	7	958	3.50 ,,	3.55	3.55	Mulgrave's Islands	Australia
25	8	959	3.51 ,,	3.56	3.56	Mull, *Island*	Hebrides
25	9	960	3.52 ,,	3.57	3.57	Mundania	Turkey in Asia
25	10	961	3.53 ,,	3.58	3.58	Mussendom, *Cape*	Arabia
25	11	962	3.54 ,,	3.59	3.59	Moen, *Island*	Denmark
25	12	963	3.55 ,,	4. 0	4. 0	Nakkehoved, *Lights*	Sealand
25	13	964	3.56 ,,	4. 1	4. 1	Nakskov	Laaland
25	14	965	3.57 ,,	4. 2	4. 2	Nanga-Saqui	Japan
25	15	966	3.58 ,,	4. 3	4. 3	Nankin	China
25	16	967	3.59 ,,	4. 4	4. 4	Nantes	France
25	17	968	4. 1 ,,	4. 5	4. 5	Nantucket, *Island*, *Light*	U. States, N. A.
25	18	969	4. 2 ,,	4. 6	4. 6	Nantucket Shoals	U. States, N. A.
25	19	970	4. 3 ,,	4. 7	4. 7	Nao, *Cape*	Spain
25	20	971	4. 4 ,,	4. 8	4. 8	Napakian	Likeio Isles
25	21	972	4. 5 ,,	4. 9	4. 9	Napoli di Malvasia	Greece
25	22	973	4. 6 ,,	4.10	4.10	Napoli di Romania	Greece
25	23	974	4. 7 ,,	4.11	4.11	Nargoe, *Light*	Gulf of Finland
25	24	975	4. 8 ,,	4.12	4.12	Narva, *Light*	Russia
25	27	976	4. 9 ,,	4.13	4.13	Nassau	Bahama Isles
25	28	977	4.10 ,,	4.14	4.14	Natal, *Cape*	Africa, SE
25	29	978	4.11 ,,	4.15	4.15	Natuna, Great, *Island*	China Sea
25	30	979	4.12 ,,	4.16	4.16	Natuna, North, *Island*	China Sea
25	31	980	4.13 ,,	4.17	4.17	Natuna, South, *Island*	China Sea
25	32	981	4.14 ,,	4.18	4.18	Naturaliste, *Cape*	New Holland
25	33	982	4.15 ,,	4.19	4.19	Navarino	Morea
25	34	983	4.16 ,,	4.20	4.20	Naxos, *Island*	Archipelago
25	35	984	4.17 ,,	4.21	4.21	Naples, *Light*	Italy
						Naze, The, see Lindersnœs	
25	36	985	4.18 ,,	4.22	4.22	Needles, *Rocks*, *Light*	England
25	37	986	4.19 ,,	4.23	4.23	Negapatam	Coast of Coromandel
25	38	987	4.20 ,,	4.24	4.24	Negrais, *Cape*	Pegu Coast
25	39	988	4.21 ,,	4.25	4.25	Negro, *Cape*	Africa, W
25	40	989	4.22 ,,	4.26	4.26	Negropont, *Island*	Archipelago

COMPOUND TABLE.

Pair Signals. No.	No.	Numerals.	Time of the Day. H. M.	Latitude South. ° ′	Longitude West. ° ′	Nations, Countries, Seas, Seaports, Capes, &c.	
26	1	990	4.23 P.M.	4.27	4.27	Nervoe-Bake	Gulf of Finland
26	2	991	4.24 „	4.28	4.28	Neustadt	Holstein
26	3	992	4.25 „	4.29	4.29	Neuwerk, *Lights*	Hanover
26	4	993	4.26 „	4.30	4.30	Neversink, *Lights*	New York
26	5	994	4.27 „	4.31	4.31	Nevis, *Island*	West Indies
26	6	995	4.28 „	4.32	4.32	Newarp, *floating Lights*	England
26	7	996	4.29 „	4.33	4.33	New Bedford	U. States, N.A.
26	8	997	4.30 „	4.34	4.34	New Bern	U. States, N.A.
26	9	998	4.31 „	4.35	4.35	New Britannia, *Islands*	Australia
26	10	999	4.32 „	4.36	4.36	New Brunswick	U. States, N.A.
26	11	1000	4.33 „	4.37	4.37	New Brunswick, (*English*)	North America
26	12	1001	4.34 „	4.38	4.38	Newberry Port	U. States, N.A.
26	13	1002	4.35 „	4.39	4.39	New Caledonia, *Island*	Australia
26	14	1003	4.36 „	4.40	4.40	Newcastle	England
26	15	1004	4.37 „	4.41	4.41	New England	U. States, N.A.
26	16	1005	4.38 „	4.42	4.42	Newfoundland	North America
26	17	1006	4.39 „	4.43	4.43	Newfoundland, Banks of	Newfoundland
26	18	1007	4.40 „	4.44	4.44	New Guinea, *Island*	Australia
26	19	1008	4.41 „	4.45	4.45	New Hampshire	U. States, N.A.
26	20	1009	4.42 „	4.46	4.46	Newhaven, *Lights*	England
26	21	1010	4.43 „	4.47	4.47	Newhaven, *Light*	U. States, N.A.
26	22	1011	4.44 „	4.48	4.48	New Hebrides, *Islands*	Australia
26	23	1012	4.45 „	4.49	4.49	New Holland, *Island*	Australia
26	24	1013	4.46 „	4.50	4.50	New London, *Light*	U. States, N.A.
26	27	1014	4.47 „	4.51	4.51	New Orleans	U. States, N.A.
26	28	1015	4.48 „	4.52	4.52	New Port	U. States, N.A.
26	29	1016	4.49 „	4.53	4.53	New South Wales	New Holland
26	30	1017	4.50 „	4.54	4.54	New York	U. States, N.A.
26	31	1018	4.51 „	4.55	4.55	New Zealand, *Islands*	Australia
26	32	1019	4.52 „	4.56	4.56	Nexoe	Bornholm
26	33	1020	4.53 „	4.57	4.57	Nicaria, *Island*	Archipelago
26	34	1021	4.54 „	4.58	4.58	Nice	France
26	35	1022	4.55 „	4.59	4.59	Nicobar, Great, *Island*	Indian Ocean
26	36	1023	4.56 „	5. 0	5. 0	Nicobar, Little, *Island*	Indian Ocean
26	37	1024	4.57 „	5. 1	5. 1	Nicobar, Car, *Island*	Indian Ocean
26	38	1025	4.58 „	5. 2	5. 2	Nicholas, St., *Island*	Cape Verd Isles
26	39	1026	4.59 „	5. 3	5. 3	Niddingen, *Island, Lights*	Cattegat
26	40	1027	5. 1 „	5. 4	5. 4	Nipon, *Island*	Japan

COMPOUND TABLE.

Pair Signals. No.	Pair Signals. No.	Numerals.	Time of the Day.	Latitude South.	Longitude West.	Nations, Countries, Seas, Seaports, Capes, &c.	
			H. M.	° ′	° ′		
27	1	1028	5. 2 P.M.	5. 5	5. 5	Niragua . . .	Panama
27	2	1029	5. 3 „	5. 6	5. 6	Noa, Pt. . .	Persia
27	3	1030	5. 4 „	5. 7	5. 7	Nootka Sound . .	America, NW
27	4	1031	5. 5 „	5. 8	5. 8	North Cape . .	Finmark, N
27	5	1032	5. 6 „	5. 9	5. 9	North East, Cape . .	Siberia
27	6	1033	5. 7 „	5.10	5.10	Nordstrand, *Island* .	Denmark, W
27	7	1034	5. 8 „	5.11	5.11	North Sea.	
27	8	1035	5. 9 „	5.12	5.12	Nore, *floating Light* . . .	England
27	9	1036	5.10 „	5.13	5.13	Norfolk, *Island* . . .	Australia
27	10	1037	5.11 „	5.14	5.14	Norrkoeping . . .	Sweden, E
27	11	1038	5.12 „	5.15	5.15	Northumberland, *Cape* .	New S. Wales
27	12	1039	5.13 „	5.16	5.16	Nova Scotia, *Island* .	North America
27	13	1040	5.14 „	5.17	5.17	Nowaja Semlia, *Islands* .	Icy Ocean
27	14	1041	5.15 „	5.18	5.18	Nun Rock . .	Orkneys, W.
27	15	1042	5.16 „	5.19	5.19	Nyborg . . .	Funen
27	16	1043	5.17 „	5.20	5.20	Ob, Gulf of . .	Siberia
27	17	1044	5.18 „	5.21	5.21	Ocean, Atlantic, North.	
27	18	1045	5.19 „	5.22	5.22	Ocean, Atlantic, South.	
27	19	1046	5.20 „	5.23	5.23	Ocean, Indian.	
27	20	1047	5.21 „	5.24	5.24	Ocean, Northern.	
27	21	1048	5.22 „	5.25	5.25	Ocean, Southern.	
27	22	1049	5.23 „	5.26	5.26	Ocean, Pacific.	
27	23	1050	5.24 „	5.27	5.27	Odensholm, *Light* .	Gulf of Finland
27	24	1051	5.25 „	5.28	5.28	Odessa, *Light* . .	Russia
27	25	1052	5.26 „	5.29	5.29	Oleron, *Island* . .	France, W
27	26	1053	5.27 „	5.30	5.30	Omba, *Island* . .	Molucca Isles, SW
27	29	1054	5.28 „	5.31	5.31	Omoe . . .	Great Belt
27	30	1055	5.29 „	5.32	5.32	Oporto . . .	Portugal
27	31	1056	5.30 „	5.33	5.33	Oran . . .	Africa, N
27	32	1057	5.31 „	5.34	5.34	Orchilla, *Island* . .	Caribbean Sea
27	33	1058	5.32 „	5.35	5.35	Orfordness, *Lights* . .	England
27	34	1059	5.33 „	5.36	5.36	Orfui, *Cape* . . .	Africa, E
27	35	1060	5.34 „	5.37	5.37	Orinoco, *River* . .	South America
27	36	1061	5.35 „	5.38	5.38	Oristano . . .	Sardinia
						Orkney Islands, see pair signals, 40.21	
27	37	1062	5.36 „	5.39	5.39	Orlando, d', *Cape* . .	Sicily
27	38	1063	5.37 „	5.40	5.40	Orlows-Islands . .	Cape Horn, S
27	39	1064	5.38 „	5.41	5.41	Ormus, *Island* . .	Persian Gulf
27	40	1065	5.39 „	5.42	5.42	Orne, l', *River*, *Lights* . .	France

COMPOUND TABLE.

Pair Signals. No.	No.	Numerals.	Time of the Day. H. M.	Latitude South. ° '	Longitude West. ° '	Nations, Countries, Seas, Seaports, Capes, &c.	
28	1	1066	5.40 PM.	5.43	5.43	ORTEGAL, *Cape*	Spain
28	2	1067	5.41 „	5.44	5.44	ORTONA	Naples
28	3	1068	5.42 „	5.45	5.45	ORTON ISLAND	North America
28	4	1069	5.43 „	5.46	5.46	OSTEND, *Lights*	Belgium
28	5	1070	5.44 „	5.47	5.47	OSTIA	Italy
28	6	1071	5.45 „	5.48	5.48	OTAHEITI, *Island*	Australia
28	7	1072	5.46 „	5.49	5.49	OUESSANT, *Island, Light*	France, W
28	8	1073	5.47 „	5.50	5.50	OUTAS, *Light*	Portugal
28	9	1074	5.48 „	5.51	5.51	OWHYEE, *Island*	Sandwich Isles
28	10	1075	5.49 „	5.52	5.52	OWERS, *floating Light*	England
28	11	1076	5.50 „	5.53	5.53	OXOE, *Light*	Norway
28	12	1077	5.51 „	5.54	5.54	PACKERORT, *Light*	Gulf of Finland
28	13	1078	5.52 „	5.55	5.55	PACKET-ROCK	St. Thomas, S
28	14	1079	5.53 „	5.56	5.56	PADARAN, *Cape*	Cochin China
28	15	1080	5.54 „	5.57	5.57	PALAMBAN	Sumatra
28	16	1081	5.55 „	5.58	5.58	PALAWAN, *Island*	Borneo, NE
28	17	1082	5.56 „	5.59	5.59	PALERMO, *Light*	Sicily
28	18	1083	5.57 „	6. 0	6. 0	PALMA, *Island*	Canary Isles
28	19	1084	5.58 „	6. 1	6. 1	PALMAS, *Cape*	Africa, W
28	20	1085	5.59 „	6. 2	6. 2	PALMERSLAND	South Atlantic Ocean
28	21	1086	6. 1 „	6. 3	6. 3	PALMYRA, PT.	Ceylon
28	22	1087	6. 2 „	6. 4	6. 4	PALMYRA, *Island*	Australia
28	23	1088	6. 3 „	6. 5	6. 5	PALOS, CAPE DE	Spain
28	24	1089	6. 4 „	6. 6	6. 6	PANAMA	South America
28	25	1090	6. 5 „	6. 7	6. 7	PANTELLARIA, *Island*	Mediterranean
28	26	1091	6. 6 „	6. 8	6. 8	PASSAGE, *Strait*	St. Thomas & Porto-Rico
28	29	1092	6. 7 „	6. 9	6. 9	PATANI, *Cape*	Gulf of Siam
28	30	1093	6. 8 „	6.10	6.10	PATERNOSTER, PT.	Africa, S W
28	31	1094	6. 9 „	6.11	6.11	PATERNOSTER, *Rocks*	Skagerak
28	32	1095	6.10 „	6.12	6.12	PATMOS, *Island*	Archipelago
28	33	1096	6.11 „	6.13	6.13	PATRAS	Morea
28	34	1097	6.12 „	6.14	6.14	PORT PATRICK, *Light*	Scotland
28	35	1098	6.13 „	6.15	6.15	PAUL, ST., *Cape*	Guinea
28	36	1099	6.14 „	6.16	6.16	PAXO, *Island, Light*	Ionian Isles
28	37	1100	6.15 „	6.17	6.17	PELAGOSA, *Island*	Adriatic
28	38	1101	6.16 „	6.18	6.18	PELEW, *Islands*	Australia
28	39	1102	6.17 „	6.19	6.19	PELWORM, *Island*	Denmark, W
28	40	1103	6.18 „	6.20	6.20	PEMBA, *Island*	Africa, E

COMPOUND TABLE.

Pair Signals. No.	No.	Numerals.	Time of the Day.	Latitude South.	Longitude West.	Nations, Countries, Seas, Seaports, Capes, &c.	
			H. M.	° ′	° ′		
29	1	1104	6.19 P.M.	6.21	6.21	Penas, *Cape*	Spain
29	2	1105	6.20 ,,	6.22	6.22	Penedo de St. Pedro, *Rock*	Atlantic
29	3	1106	6.21 ,,	6.23	6.23	Peniscola	Spain
29	4	1107	6.22 ,,	6.24	6.24	Penmarch, Pt. de, *Light*	France
29	5	1108	6.23 ,,	6.25	6.25	Penobscot	U. States, N. A.
29	6	1109	6.24 ,,	6.26	6.26	Pensacola, *Light*	U. States, N. A.
29	7	1110	6.25 ,,	6.27	6.27	Pentland Firth	Scotland
29	8	1111	6.26 ,,	6.28	6.28	Pentland-Skerries, *Lights*	Scotland
29	9	1112	6.27 ,,	6.29	6.29	Penzance, *Light*	England
29	10	1113	6.28 ,,	6.30	6.30	Pernambuco	Brazil
29	11	1114	6.29 ,,	6.31	6.31	Pernau	Lifland
29	12	1115	6.30 ,,	6.32	6.32	Persaim	Pegu Coast
29	13	1116	6.31 ,,	6.33	6.33	Persian Gulf	Persia
29	14	1117	6.32 ,,	6.34	6.34	Pescadores, *Islands*	Australia
29	15	1118	6.33 ,,	6.35	6.35	Peterhead	Scotland
29	16	1119	6.34 ,,	6.36	6.36	Peter and Paul's Harbour	Kamtschatka
29	17	1120	6.35 ,,	6.37	6.37	Petersburg, St.	Russia
29	18	1121	6.36 ,,	6.38	6.38	Philadelphia	U. States, N. A.
29	19	1122	3.37 ,,	6.39	6.39	Philippine Islands	Asia, E
29	20	1123	6.38 ,,	6.40	6.40	Pico, *Island*	Azores
29	21	1124	6.39 ,,	6.41	6.41	Pierre, St.	Martinique
29	22	1125	6.40 ,,	6.42	6.42	Pilier, *Island, Light*	France, W
29	23	1126	6.41 ,,	6.43	6.43	Pillau, *Light*	Prussia
29	24	1127	6.42 ,,	6.44	6.44	Piombino	Italy
29	25	1128	6.43 ,,	6.45	6.45	Pirate's Island	Tung guin Bay
29	26	1129	6.44 ,,	6.46	6.46	Pisa	Italy
29	27	1130	6.45 ,,	6.47	6.47	Pitea	Sweden
29	28	1131	6.46 ,,	6.48	6.48	Pitcairn, *Island*	Australia
29	31	1132	6.47 ,,	6.49	6.49	Planier, *Island, Light*	France, S
29	32	1133	6.48 ,,	6.50	6.50	Plata, Rio de la	South America
29	33	1134	6.49 ,,	6.51	6.51	Plettenburg Bay	Africa, S
29	34	1135	6.50 ,,	6.52	6.52	Plymouth, *floating Lights*	England
29	35	1136	6.51 ,,	6.53	6.53	Pola	Istria
29	36	1137	6.52 ,,	6.54	6.54	Polder	Dunkirk, NW
29	37	1138	6.53 ,,	6.55	6.55	Pomona, *Island*	Orkneys
29	38	1139	6.54 ,,	6.56	6.56	Pondicherry	Coast of Coromandel
29	39	1140	6.55 ,,	6.57	6.57	Ponte Vechio	Corsica
29	40	1141	6.56 ,,	6.58	6.58	Pontiana	Borneo

COMPOUND TABLE.

Pair Signals, No	No.	Numerals.	Time of the Day.	Latitude South.	Longitude West.	Nations, Countries, Seas, Seaports, Capes, &c.	
			H. M.	° ′	° ′		
30	1	1142	6.57 P.M.	6.59	6.59	Ponza, *Island*	Pontic Islands
30	2	1143	6.58 „	7. 0	7. 0	Port au Prince	Hayti
30	3	1144	6.59 „	7. 1	7. 1	Portland, *Light*	U. States, N. A.
30	4	1145	7. 1 „	7. 2	7. 2	Portland-Bill, *Lights*	England
30	5	1146	7. 2 „	7. 3	7. 3	Porto-Bello	South America
30	6	1147	7. 3 „	7. 4	7. 4	Porto Caballo	South America
30	7	1148	7. 4 „	7. 5	7. 5	Porto Praja	Cape Verd Isles
30	8	1149	7. 5 „	7. 6	7. 6	Porto-Rico, *Island*	West Indies
30	9	1150	7. 6 „	7. 7	7. 7	Porto-Santo, *Island*	Madeira, NE
30	10	1151	7. 7 „	7. 8	7. 8	Port Royal	Jamaica
30	11	1152	7. 8 „	7. 9	7. 9	Portsmouth, *Light*	U. States, N. A.
30	12	1153	7. 9 „	7.10	7.10	Portsmouth	England
30	13	1154	7.10 „	7.11	7.11	Powell's Island	South Atlantic
30	14	1155	7.11 „	7.12	7.12	Preparis Island	Bay of Bengal
30	15	1156	7.12 „	7.13	7.13	Prince Edward Islands	Indian Ocean
30	16	1157	7.13 „	7.14	7.14	Prince of Wales Cape	America, NW
30	17	1158	7.14 „	7.15	7.15	Prince's Islands	Africa, W
30	18	1159	7.15 „	7.16	7.16	Providence, *Island*	Bahama Isles
30	19	1160	7.16 „	7.17	7.17	Providence	Pacific Ocean
30	20	1161	7.17 „	7.18	7.18	Providence	U. States, N. A.
30	21	1162	7.18 „	7.19	7.19	Prœstoe	Sealand
30	22	1163	7.19 „	7.20	7.20	Pulo Penang	Straits of Malacca
30	23	1164	7.20 „	7.21	7.21	Quebec	North America
30	24	1165	7.21 „	7.22	7.22	Querqueville, *Light*	France
30	25	1166	7.22 „	7.23	7.23	Quiberon	France
30	26	1167	7.23 „	7.24	7.24	Quilleboeuf, *Light*	France
30	27	1168	7.24 „	7.25	7.25	Quiloa	Africa, E
30	28	1169	7.25 „	7.26	7.26	Race, *Cape*, *Light*	Newfoundland
30	31	1170	7.26 „	7.27	7.27	Radacks Islands	Australia
30	32	1171	7.27 „	7.28	7.28	Ragusa	Dalmatia
30	33	1172	7.28 „	7.29	7.29	Ram-Head	England
30	34	1173	7.29 „	7.30	7.30	Ramisseram, *Island*	Gulf of Manar
30	35	1174	7.30 „	7.31	7.31	Ramsgate, *Light*	England
30	36	1175	7.31 „	7.32	7.32	Ramus, *Cape*	Canara
30	37	1176	7.32 „	7.33	7.33	Rangoon	East Indies
30	38	1177	7.33 „	7.34	7.34	Ras el Kire	Africa, E
30	39	1178	7.34 „	7.35	7.35	Ras el Reccan	Arabian Gulf
30	40	1179	7.35 „	7.36	7.36	Raumo	Finland

COMPOUND TABLE.

Pair Signals No.	Pair Signals No.	Numerals.	Time of the Day.	Latitude South.	Longitude West.	Nations, Countries, Seas, Seaports, Capes, &c.	
			H. M.	° ′	° ′		
31	1	1180	7.36 P.M.	7.37	7.37	Ravenna, *Light*	Italy
31	2	1181	7.37 „	7.38	7.38	Recherches, *Archipelago*	New Holland
31	3	1182	7.38 „	7.39	7.39	Reculver	England
31	4	1183	7.39 „	7.40	7.40	Redang, *Island*	Gulf of Siam
31	5	1184	7.40 „	7.41	7.41	Redondo, *Island*	West Indies
31	6	1185	7.41 „	7.42	7.42	Re, Isle de, *Light*	France, W
31	7	1186	7.42 „	7.43	7.43	Reggio	Calabria
31	8	1187	7.43 „	7.44	7.44	Reikavig	Iceland
31	9	1188	7.44 „	7.45	7.45	Resterhooft, *Light*	Baltic
31	10	1189	7.45 „	7.46	7.46	Resolution, *Island*	Hudson's Straits
31	11	1190	7.46 „	7.47	7.47	Reval	Russia
31	12	1191	7.47 „	7.48	7.48	Revilla Gigedo, *Islands*	California, SW
31	13	1192	7.48 „	7.49	7.49	Revsnœs	Great Belt
31	14	1193	7.49 „	7.50	7.50	Rhode-Island	U. States, N. A.
31	15	1194	7.50 „	7.51	7.51	Rhodos, *Island, Light*	Mediterranean
31	16	1195	7.51 „	7.52	7.52	Rhone, *River*	France
31	17	1196	7.52 „	7.53	7.53	Ribe	Jutland
31	18	1197	7.53 „	7.54	7.54	Richmond	U. States, N. A.
31	19	1198	7.54 „	7.55	7.55	Ridge	Channel
31	20	1199	7.55 „	7.56	7.56	Riga, *Lights*	Russia
31	21	1200	7.56 „	7.57	7.57	Rizzuto, *Cape*	Italy, S
31	22	1201	7.57 „	7.58	7.58	Rober-Hooft, *Light*	Pomerania
31	23	1202	7.58 „	7.59	7.59	Robert-Knude	Jutland, NW
31	24	1203	7.59 „	8. 0	8. 0	Roca, Cape de, *Light*	Portugal
31	25	1204	8. 1 „	8. 1	8. 1	Roccas, *Islands*	South America
31	26	1205	8. 2 „	8. 2	8. 2	Rochefort	France
31	27	1206	8. 3 „	8. 3	8. 3	Rochelle, *Light*	France
31	28	1207	8. 4 „	8. 4	8. 4	Roches Douvres	Guernsey, S
31	29	1208	8. 5 „	8. 5	8. 5	Roderiquez, *Island*	Indian Ocean
31	30	1209	8. 6 „	8. 6	8. 6	Rodney, Port	America, NW
31	33	1210	8. 7 „	8. 7	8. 7	Rokol, *Reef*	Hebrides
31	34	1211	8. 8 „	8. 8	8. 8	Romania, *Cape*	Malacca, S
31	35	1212	8. 9 „	8. 9	8. 9	Romanzow, *Island*	Radacks Islands
31	36	1213	8.10 „	8.10	8.10	Romsoe, *Island*	Great Belt
31	37	1214	8.11 „	8.11	8.11	Romoe, *Island*	Denmark, W
31	38	1215	8.12 „	8.12	8.12	Rona, *Rock*	Orkneys, W
31	39	1216	8.13 „	8.13	8.13	Ronaldshay, North	Orkneys
31	40	1217	8.14 „	8.14	8.14	Ronaldshay, South	Orkneys

COMPOUND TABLE.

Pair Signals. No.	No.	Numerals.	Time of the Day.	Latitude South.	Longitude West.	Nations, Countries, Seas, Seaports, Capes, &c.	
			H. M.	° ′	° ′		
32	1	1218	8.15 P.M.	8.15	8.15	Ronas Hill	Shetland Isles
32	2	1219	8.16 ,,	8.16	8.16	Rondoe, *Light*	Norway, W
32	3	1220	8.17 ,,	8.17	8.17	Rosetta	Egypt
32	4	1221	8.18 ,,	8.18	8.18	Rostock	Mecklenburg
32	5	1222	8.19 ,,	8.19	8.19	Rota	Spain
32	6	1223	8.20 ,,	8.20	8.20	Rothsay	Scotland
32	7	1224	8.21 ,,	8.21	8.21	Rotterdam	Holland
32	8	1225	8.22 ,,	8.22	8.22	Rouen	France
32	9	1226	8.23 ,,	8.23	8.23	Rovigno	Istria
32	10	1227	8.24 ,,	8.24	8.24	Runs, Point of, *Light*	Scotland, W
32	11	1228	8.25 ,,	8.25	8.25	Runoe, *Light*	Gulf of Riga
32	12	1229	8.26 ,,	8.26	8.26	Rye, *Lights*	England
32	13	1230	8.27 ,,	8.27	8.27	Rugen, *Island*	Baltic
32	14	1231	8.28 ,,	8.28	8.28	Rugenwalde	Pomerania
32	15	1232	8.29 ,,	8.29	8.29	Red Sea	Arabia
32	16	1233	8.30 ,,	8.30	8.30	Roedskœr, *Light*	Gulf of Finland
32	17	1234	8.31 ,,	8.31	8.31	Roenne	Bornholm
32	18	1235	8.32 ,,	8.32	8.32	Roenskœr, *Light*	Gulf of Finland
32	19	1236	8.33 ,,	8.33	8.33	Saba, *Island*	West Indies
32	20	1237	8.34 ,,	8.34	8.34	Sables d'Olonne, *Lights*	France
32	21	1238	8.35 ,,	8.35	8.35	Sable, *Cape*	Nova Scotia
32	22	1239	8.36 ,,	8.36	8.36	Said	Syria
32	23	1240	8.37 ,,	8.37	8.37	Saigon	Anam
32	24	1241	8.38 ,,	8.38	8.38	Saintes	Bay of Biscay
32	25	1242	8.39 ,,	8.39	8.39	Sal, *Island*	Cape Verd Isles
32	26	1243	8.40 ,,	8.40	8.40	Salanga, *or* Junkseylon	Malay Coast
32	27	1244	8.41 ,,	8.41	8.41	Salayer, *Island*	Celebes, E
32	28	1245	8.42 ,,	8.42	8.42	Saldanha Bay	Africa, S
32	29	1246	8.43 ,,	8.43	8.43	Salem, *Lights*	U. States, N.A.
32	30	1247	8.44 ,,	8.44	8.44	Salerno	Naples
32	33	1248	8.45 ,,	8.45	8.45	Salina, *Island*	Sicily, N
32	34	1249	8.46 ,,	8.46	8.46	Solomon's Islands	Australia
32	35	1250	8.47 ,,	8.47	8.47	Salonica, *Light*	Turkey
32	36	1251	8.48 ,,	8.48	8.48	Saloweskoi, *Island*	White Sea
32	37	1252	8.49 ,,	8.49	8.49	Saltholm	The Sound
32	38	1253	8.50 ,,	8.50	8.50	Salvador, St., *Island*	West Indies
32	39	1254	8.51 ,,	8.51	8.51	Salvages, *Islands*	Canary Isles, N
32	40	1255	8.52 ,,	8.52	8.52	Samar, *Island*	Philippine Isles

COMPOUND TABLE.

Pair Signals No.	Pair Signals No.	Numerals.	Time of the Day.	Latitude South.	Longitude West.	Nations, Countries, Seas, Seaports, Capes, &c.	
			H. M.	° ′	° ′		
33	1	1256	8.53 PM.	8.53	8.53	Samarang	Java
33	2	1257	8.54 „	8.54	8.54	Samos, *Island*	Archipelago
33	3	1258	8.55 „	8.55	8.55	Samsoe, *Island*	Cattegat
33	4	1259	8.56 „	8.56	8.56	Sandefiord	Norway
33	5	1260	8.57 „	8.57	8.57	Sandelbosch, *Island*	Timor, S
33	6	1261	8.58 „	8.58	8.58	Sandy, *Cape*	N. S. Wales
33	7	1262	8.59 „	8.59	8.59	Sandyhook, *Light*	U. States, N.A.
33	8	1263	9. 1 „	9. 0	9. 0	Sandoe	Feroe Isles
33	9	1264	9. 2 „	9. 1	9. 1	Sandwich-Land	South Atlantic Ocean
33	10	1265	9. 3 „	9. 2	9. 2	Sandwich-Islands	Australia
33	11	1266	9. 4 „	9. 3	9. 3	Sangier, *Island*	Celebes, N
33	12	1267	9. 5 „	9. 4	9. 4	Santander	Spain
33	13	1268	9. 6 „	9. 5	9. 5	Santorin, *Island*	Archipelago
33	14	1269	9. 7 „	9. 6	9. 6	Savannah	U. States, N.A.
33	15	1270	9. 8 „	9. 7	9. 7	Saya de Malha, *Bank*	Indian Ocean
33	16	1271	9. 9 „	9. 8	9. 8	Scarborough, *Light*	England
33	17	1272	9.10 „	9. 9	9. 9	Scarpanto, *Island*	Mediterranean
33	18	1273	9.11 „	9.10	9.10	Schelde, *River*	Holland
33	19	1274	9.12 „	9.11	9.11	Scheveningen, *Light*	Holland
33	20	1275	9.13 „	9.12	9.12	Scilly, *Islands*	England, SW
33	21	1276	9.14 „	9.13	9.13	Sleswick	Denmark
33	22	1277	9.15 „	9.14	9.14	Schowen, *Lights*	Holland
33	23	1278	9.16 „	9.15	9.15	Scio, *Island*	Archipelago
33	24	1279	9.17 „	9.16	9.16	Scotland	Great Britain
33	25	1280	9.18 „	9.17	9.17	Sebastian, St., *Cape*, *Light*	Cadiz
33	26	1281	9.19 „	9.18	9.18	Sebastian, St., *Light*	Bilboa, E
33	27	1282	9.20 „	9.19	9.19	Sebastian's (St.) Bay	Africa, S
33	28	1283	9.21 „	9.20	9.20	Seine, *River*	France
33	29	1284	9.22 „	9.21	9.21	Seieroe, *Island*	Great Belt
33	30	1285	9.23 „	9.22	9.22	Senegal, *River*	Africa, W
33	31	1286	9.24 „	9.23	9.23	Senegambia	Africa, W
33	32	1287	9.25 „	9.24	9.24	Senjen, *Island*	Finmark
33	35	1288	9.26 „	9.25	9.25	Serampore (Frederiksnagor)	Bengal
33	36	1289	9.27 „	9.26	9.26	Sark, *Island*	Channel
33	37	1290	9.28 „	9.27	9.27	Seskœr, *Light*	Gulf of Finland
33	38	1291	9.29 „	9.28	9.28	Setuval, or St. Ubes	Portugal
33	39	1292	9.30 „	9.29	9.29	Sevastopol	Crimea
33	40	1293	9.31 „	9.30	9.30	Shannon, *River*	Ireland

COMPOUND TABLE.

Pair Signals. No.	Pair Signals. No.	Numerals.	Time of the Day.	Latitude South.	Longitude West.	Nations, Countries, Seas, Seaports, Capes, &c.
			H.M.	° ′	° ′	
34	1	1294	9.32 P.M.	9.31	9.31	SHETLAND ISLES . Scotland, N
34	2	1295	9.33 „	9.32	9.32	SHETLAND, SOUTH . South Atlantic Ocean
34	3	1296	9.34 „	9.33	9.33	SHIELDS, *Lights* . . England
34	4	1297	9.35 „	9.34	9.34	SHOREHAM, *Lights* . . England
34	5	1298	9.36 „	9.35	9.35	SIAM, GULF OF . . China Sea
34	6	1299	9.37 „	9.36	9.36	SIBERIA, NEW, *Island* . North Icy Ocean
34	7	1300	9.38 „	9.37	9.37	SICILY, *Island* . . Mediterranean
34	8	1301	9.39 „	9.38	9.38	SIERRA-LEONE . . Africa, W
34	9	1302	9.40 „	9.39	9.39	SINGAPORE . . . Malacca
34	10	1303	9.41 „	9.40	9.40	SEALAND, *Island* . . Denmark
34	11	1304	9.42 „	9.41	9.41	SEALAND'S REEF . Sealand, NW
34	12	1305	9.43 „	9.42	9.42	SKAW, *Light* . . Jutland, N
34	13	1306	9.44 „	9.43	9.43	SKAGERAK . Jutland, Norway, & Sweden
34	14	1307	9.45 „	9.44	9.44	SKALHOLT . . . Iceland
34	15	1308	9.46 „	9.45	9.45	SKELLIGS, *Lights* . . . Ireland
34	16	1309	9.47 „	9.46	9.46	SKIELSKIOR . . . Sealand
34	17	1310	9.48 „	9.47	9.47	SKUDDESNŒS, *Light* . . Karmoe
34	18	1311	9.49 „	9.48	9.48	SKY, *Island* . . Hebrides
34	19	1312	9.50 „	9.49	9.49	SKYRO, *Island* . . Archipelago
34	20	1313	9.51 „	9.50	9.50	SMALLS, *Light* . St. George's Channel
34	21	1314	9.52 „	9.51	9.51	SMYRNA . . Turkey in Asia
34	22	1315	9.53 „	9.52	9.52	SMOELEN, *Island* . . Norway, W
34	23	1316	9.54 „	9.53	9.53	SOCIETY-ISLANDS . . Australia
34	24	1317	9.55 „	9.54	9.54	SOCOTORA, *Island* . . Arabian Sea
34	25	1318	9.56 „	9.55	9.55	SOFALA . . . Africa, E
34	26	1319	9.57 „	9.56	9.56	SOMBRERO, *Island* . West Indies
34	27	1320	9.58 „	9.57	9.57	SOMMEROE, *Light* . Gulf of Finland
34	28	1321	9.59 „	9.58	9.58	SORRENTO . . . Italy
34	29	1322	10. 1 „	9.59	9.59	SEA, BLACK . . Russia & Turkey
						SOUND, THE, see pair signals, 40.20.
34	30	1323	10. 2 „	10. 0	10. 0	SOUTHAMPTON . . England
34	31	1324	10. 3 „	10. 1	10. 1	SPALATRO . . Dalmatia
34	32	1325	10. 4 „	10. 2	10. 2	SPANISH-SEA . . Portugal, W
34	35	1326	10. 5 „	10. 3	10. 3	SPARŒ BAKE . . Sweden
34	36	1327	10. 6 „	10. 4	10. 4	SPARTEL, *Cape* . . . Morocco
34	37	1328	10. 7 „	10. 5	10. 5	SPENCER'S, or BONAPARTE'S BAY New S. Wales
34	38	1329	10. 8 „	10. 6	10. 6	SPEZIA . . . Sardinia
34	39	1330	10. 9 „	10. 7	10. 7	SPEZZIA, *Island* . Gulf of Napoli
34	40	1331	10.10 „	10. 8	10. 8	SPITHEAD . . England

COMPOUND TABLE.

Pair Signals. No.	No.	Numerals.	Time of the Day.	Latitude South.	Longitude West.	Nations, Countries, Seas, Seaports, Capes, &c.	
			H. M.	° ′	° ′		
35	1	1332	10.11 PM.	10. 9	10. 9	Spitsbergen	North Icy Ocean
35	2	1333	10.12 „	10.10	10.10	Sprogoe, *Light*	Great Belt
35	3	1334	10.13 „	10.11	10.11	Spurn-Head, *Lights*	England
35	4	1335	10.14 „	10.12	10.12	Spurn, *floating Light*	England
35	5	1336	10.15 „	10.13	10.13	Staffa, *Island*	Hebrides
35	6	1337	10.16 „	10.14	10.14	Stalimene, *Island*	Archipelago
35	7	1338	10.17 „	10.15	10.15	Stanchio, *Island*	Archipelago
35	8	1339	10.18 „	10.16	10.16	Stanford-Gat, *floating Lights*	England
35	9	1340	10.19 „	10.17	10.17	Stangskœr-Bake	Bay of Norrkoeping
35	10	1341	10.20 „	10.18	10.18	Staples, *Light*	England
35	11	1342	10.21 „	10.19	10.19	Start Point	England
35	12	1343	10.22 „	10.20	10.20	Start, *Light*	Orkneys
35	13	1344	10.23 „	10.21	10.21	Stat, *Cape*	Norway, W
35	14	1345	10.24 „	10.22	10.22	Statenhuk	Greenland, S
35	15	1346	10.25 „	10.23	10.23	Stavanger	Norway
35	16	1347	10.26 „	10.24	10.24	Stege	Moen
35	17	1348	10.27 „	10.25	10.25	Stettin	Prussia
35	18	1349	10.28 „	10.26	10.26	Stevns Klint, *Light*	Sealand
35	19	1350	10.29 „	10.27	10.27	Stockholm	Sweden
35	20	1351	10.30 „	10.28	10.28	Stolpemunde	Pomerania
35	21	1352	10.31 „	10.29	10.29	Stromboli, *Island*	Sicily
35	22	1353	10.32 „	10.30	10.30	Stromness	Orkneys
35	23	1354	10.33 „	10.31	10.31	Stroemstad	Sweden
35	24	1355	10.34 „	10.32	10.32	Stroemoe	Feroe Isles
35	25	1356	10.35 „	10.33	10.33	Suderoe	Feroe Isles
35	26	1357	10.36 „	10.34	10.34	Suez	Egypt
35	27	1358	10.37 „	10.35	10.35	Suluh, *Islands*	Borneo, NE
35	28	1359	10.38 „	10.36	10.36	Sumatra, *Island*	Indian Ocean
35	29	1360	10.39 „	10.37	10.37	Sumbo-Munk, *Rock*	Feroe Isles, S
35	30	1361	10.40 „	10.38	10.38	Sumburg-Head, *Light*	Shetland Isles, S
35	31	1362	10.41 „	10.39	10.39	Sunda, Straits of	Indian Ocean
35	32	1363	10.42 „	10.40	10.40	Sunderland, *Lights*	England
35	33	1364	10.43 „	10.41	10.41	Sundsvall	Sweden
35	34	1365	10.44 „	10.42	10.42	Sunk, *floating Light*	England, E
35	37	1366	10.45 „	10.43	10.43	Surate	Hindostan
35	38	1367	10.46 „	10.44	10.44	Surepodde, *Light*	Gulf of Finland
35	39	1368	10.47 „	10.45	10.45	Surinam	South America
35	40	1369	10.48 „	10.46	10.46	Susa	Tunis, SE

COMPOUND TABLE.

Pair Signals. No.	No.	Numerals.	Time of the Day. H. M.	Latitude South. ° ′	Longitude West. ° ′	Nations, Countries, Seas, Seaports, Capes, &c.	
36	1	1370	10.49 PM.	10.47	10.47	Svarfverort, *Light*	Oesel
36	2	1371	10.50 „	10.48	10.48	Svartklubb, *Light*	Sea of Aland
36	3	1372	10.51 „	10.49	10.49	Svendborg	Funen
36	4	1373	10.52 „	10.50	10.50	Svenskahoegar	Baltic
36	5	1374	10.53 „	10.51	10.51	Svinemunde, *Light*	Wollin
36	6	1375	10.54 „	10.52	10.52	Southern Archipelago	Australia
36	7	1376	10.55 „	10.53	10.53	Sylt, *Island*	Denmark, W
36	8	1377	10.56 „	10.54	10.54	Syra, *Island*	Archipelago
36	9	1378	10.57 „	10.55	10.55	Syracuse	Sicily
36	10	1379	10.58 „	10.56	10.56	Syria	Asia
36	11	1380	10.59 „	10.57	10.57	Soederarms-Bake	Sea of Aland
36	12	1381	11. 1 „	10.58	10.58	Soederhamn	Sweden
36	13	1382	11. 2 „	10.59	10.59	Soederkoeping	Sweden
36	14	1383	11. 3 „	11. 0	11. 0	Soelfwitsborg	Sweden
36	15	1384	11. 4 „	11. 1	11. 1	Soenderborg	Als
36	16	1385	11. 5 „	11. 2	11. 2	Soeroe	Finmark
36	17	1386	11. 6 „	11. 3	11. 3	Taasinge, *Island*	Funen, S
36	18	1387	11. 7 „	11. 4	11. 4	Table-Bay	Africa, S
36	19	1388	11. 8 „	11. 5	11. 5	Tambelan, *Island*	China Sea
36	20	1389	11. 9 „	11. 6	11. 6	Tampico	Gulf of Mexico
36	21	1390	11.10 „	11. 7	11. 7	Tanasserim	Malacca
36	22	1391	11.11 „	11. 8	11. 8	Tangier	Morocco
36	23	1392	11.12 „	11. 9	11. 9	Taormina	Sicily
36	24	1393	11.13 „	11.10	11.10	Tarbetness, *Light*	Scotland
36	25	1394	11.14 „	11.11	11.11	Tarento	Italy
36	26	1395	11.15 „	11.12	11.12	Tarifa, *Light*	Spain
36	27	1396	11.16 „	11.13	11.13	Tarragona	Spain
36	28	1397	11.17 „	11.14	11.14	Teignmouth	Persian Gulf
36	29	1398	11.18 „	11.15	11.15	Tenedos, *Island*	Archipelago
36	30	1399	11.19 „	11.16	11.16	Teneriff, *Island*	Canary Isles
36	31	1400	11.20 „	11.17	11.17	Tenez, *Cape*	Africa, N
36	32	1401	11.21 „	11.18	11.18	Terceira, *Island*	Azores
36	33	1402	11.22 „	11.19	11.19	Termingen, *Light*	Hitteren
36	34	1403	11.23 „	11.20	11.20	Termini	Sicily
36	37	1404	11.24 „	11.21	11.21	Termoli	Italy
36	38	1405	11.25 „	11.22	11.22	Tarracina	Italy
36	39	1406	11.26 „	11.23	11.23	Terranuova	Sicily
36	40	1407	11.27 „	11.24	11.24	Terschelling, *Lights*	Holland

COMPOUND TABLE.

Pair Signals. No.	Pair Signals. No.	Numerals.	Time of the Day. H. M.	Latitude South. ° ′	Longitude West. ° ′	Nations, Countries, Seas, Seaports, Capes, &c.	
37	1	1408	11.28 P.M.	11.25	11.25	Tetuan . . .	Morocco
37	2	1409	11.29 „	11.26	11.26	Texel, *Island* . . .	Holland
37	3	1410	11.30 „	11.27	11.27	Thames, *River* . .	England
37	4	1411	11.31 „	11.28	11.28	Theaki, *Island* . .	Ionian Isles
37	5	1412	11.32 „	11.29	11.29	Thomas, St., *Island* .	West Indies
37	6	1413	11.33 „	11.30	11.30	Thomas, St., *Island* . .	Africa, W
37	7	1414	11.34 „	11.31	11.31	Thorshavn . . .	Stroemoe
37	8	1415	11.35 „	11.32	11.32	Thunoe, *Light* . . .	Cattegat
37	9	1416	11.36 „	11.33	11.33	Tiburon, *Cape* . . .	Hayti
37	10	1417	11.37 „	11.34	11.34	Terra del Fuego, *Island* .	America, S
37	11	1418	11.38 „	11.35	11.35	Timoan, *Pulo*, . .	China Sea
37	12	1419	11.39 „	11.36	11.36	Timor, *Island* . . .	Australia
37	13	1420	11.40 „	11.37	11.37	Tinmouth, *Light* . .	Scotland
37	14	1421	11.41 „	11.38	11.38	Tislerne, *Islands* . .	Cattegat
37	15	1422	11.42 „	11.39	11.39	Tiornoe . . .	Sweden
37	16	1423	11.43 „	11.40	11.40	Tobago, *Island* . .	West Indies
37	17	1424	11.44 „	11.41	11.41	Tollbaken, *Light* .	Gulf of Finland
37	18	1425	11.45 „	11.42	11.42	Tongataboo, *Island* .	. Friendly Isles
37	19	1426	11.46 „	11.43	11.43	Torbay	England
37	20	1427	11.47 „	11.44	11.44	Tornea . . .	Finland
37	21	1428	11.48 „	11.45	11.45	Torres Straits . .	New Guinea
37	22	1429	11.49 „	11.46	11.46	Tortola, *Island* . .	West Indies
37	23	1430	11.50 „	11.47	11.47	Tortosa, *Cape* . . .	Spain
37	24	1431	11.51 „	11.48	11.48	Toulon	France
37	25	1432	11.52 „	11.49	11.49	Touques, *Lights* . .	France
37	26	1433	11.53 „	11.50	11.50	Trafalgar, *Cape* . .	Spain
37	27	1434	11.54 „	11.51	11.51	Tranquebar .	Coast of Coromandel
37	28	1435	11.55 „	11.52	11.52	Trapani . . .	Sicily
37	29	1436	11.56 „	11.53	11.53	Travemunde, *Light* . .	Lubec
37	30	1437	11.57 „	11.54	11.54	Treport, *Light* . .	France
37	31	1438	11.58 „	11.55	11.55	Triest	Austria
37	32	1439	11.59 „	11.56	11.56	Trincomale . . .	Ceylon
37	33	1440	———	11.57	11.57	Trindelen, *floating Light* . .	Cattegat
37	34	1441	———	11.58	11.58	Trinidad, *Island* . .	West Indies
37	35	1442	———	11.59	11.59	Tripolis . . .	Africa, N
37	36	1443	———	——	——	Tristan d'Acunha, *Island* .	Atlantic
37	39	1444	———	——	——	Tromsoe	Finmark
37	40	1445	———	——	——	Trondhiem . . .	Norway

COMPOUND TABLE.

Pair Signals. No.	Pair Signals. No.	Numerals.	Supplementary Degrees. Latitude North. ° ′	Supplementary Degrees. Latitude South. ° ′	Nations, Countries, Seas, Seaports, Capes, &c.	
38	1	1446	12.0	——	Truxillo . . .	America, W
38	2	1447	24.0	——	Tung-guin Bay . . .	China Sea
38	3	1448	36.0	——	Tunis	Africa, N
38	4	1449	48.0	——	Turk's Island . . .	West Indies
38	5	1450	60.0	——	Turon, *Cape* . .	Cochin China
38	6	1451	72.0	——	Tuskar, *revolving Light* .	St. George's Channel
38	7	1452	84.0	——	Twer	Russia
38	8	1453	——	12.0	Tyter's Island, Great .	Gulf of Finland
38	9	1454	——	24.0	Tyter's Island, Little .	Gulf of Finland
38	10	1455	——	36.0	Tonningen . . .	Sleswick
38	11	1456	——	48.0	Toensberg	Norway
38	12	1457	——	60.0	Udevalla . . .	Sweden
38	13	1458	——	72.0	Udsire, *Island* . .	Karmoe, W
38	14	1459	——	84.0	Uist, North, *Island* . . .	Hebrides
38	15	1460	——	——	Uist, South, *Island* . .	Hebrides
38	16	1461	——	——	Uleaborg . . .	Finland
38	17	1462	——	——	Ulvshale, *Passage* . . .	Moen
38	18	1463	——	——	Umea	Sweden
38	19	1464	——	——	Usedom, *Island* . .	Pomerania
38	20	1465	——	——	Ustica, *Island* . . .	Sicily, N
38	21	1466	——	——	Utklipporne . . .	Baltic
38	22	1467	——	——	Uto, *Light* . .	Gulf of Finland
38	23	1468	——	——	Utrecht	Holland
38	24	1469	——	——	Vada, *Cape* . . .	Africa, N
38	25	1470	——	——	Valderhoug, *Light* . .	Norway
38	26	1471	——	——	Valdivia . . .	America, SW
38	27	1472	——	——	Valencia . . .	Spain
38	28	1473	——	——	Vallery en Caux, St., *Light* .	France
38	29	1474	——	——	Valona . . .	Albania
38	30	1475	——	——	Valparaiso . .	America, SW
38	31	1476	——	——	Van Diemen's Land . .	New Holland
38	32	1477	——	——	Varne	Channel
38	33	1478	——	——	Venice	Italy
38	34	1479	——	——	Vera-Cruz, *Light* .	Gulf of Mexico
38	35	1480	——	——	Verd, Cape de, . . .	Africa, W
38	36	1481	——	——	Verd, Cape de, *Islands* . .	Africa, W
38	39	1482	——	——	Vestmannoe	Iceland
38	40	1483	——	——	Vesuvius	Naples

COMPOUND TABLE.

Pair Signals. No.	Pair Signals. No.	Numerals.	Supplementary Degrees. Longitude East. ° ′	Supplementary Degrees. Longitude West. ° ′	Nations, Countries, Seas, Seaports, Capes, &c.	
39	1	1484	12.0	——	Viareggio	Lucca
39	2	1485	24.0	——	Vidal, *Cape*	Africa
39	3	1486	36.0	——	Vigo	Spain
39	4	1487	48.0	——	Vincent, St., *Cape*	Portugal
39	5	1488	60.0	——	Vincent, St., *Island*	West Indies
39	6	1489	72.0	——	Virgile-Rocks	Banks of Newfoundland
39	7	1490	84.0	——	Virgin Gorda, *Island*	West Indies
					Vistula, *River*, see Weichsel	
39	8	1491	96.0	——	Vito, St., *Cape*	Sicily, NW
39	9	1492	108.0	——	Vlieland, *Light*	Holland
39	10	1493	120.0	——	Vliessing, *Light*	Holland
39	11	1494	132.0	——	Vogel-Sand	Mouth of the Elbe
39	12	1495	144.0	——	Voltas, *Cape*	Africa, W
39	13	1496	156.0	——	Voorne, *Lights*	Holland
39	14	1497	168.0	——	Vordingborg	Sealand
39	15	1498	——	12.0	Vresen	Great Belt
39	16	1499	——	24.0	Waigat's Islands	Nowaja Semlia, S
39	17	1500	——	36.0	Walcheren, *Westkappel Light*	Holland
39	18	1501	——	48.0	Warberg	Sweden
39	19	1502	——	60.0	Wardoehuus	Finmark
39	20	1503	——	72.0	Wasa	Finland
39	21	1504	——	84.0	Washington	U. States, N. A.
39	22	1505	——	96.0	Waterford-Hook, *Light*	Ireland
39	23	1506	——	108.0	Weichsel, *River*	Dantzic
39	24	1507	——	120.0	Weser, *River*	Bremen
39	25	1508	——	132.0	Weser, *floating Light*	Bremen
39	26	1509	——	144.0	Westerwick	Sweden
39	27	1510	——	156.0	Wexford	Ireland
39	28	1511	——	168.0	Whitby, *Light*	England
					White Sea, see Pair Signals, 18.36	
39	29	1512	——	——	Wiborg	Finland
39	30	1513	——	——	Wight, Isle of	Channel
39	31	1514	——	——	Wilmington	U. States, N. A.
39	32	1515	——	——	Wilson, *Cape*	New Holland
39	33	1516	——	——	Windau	Courland
39	34	1517	——	——	Wingoe	Cattegat
39	35	1518	——	——	Winterton, *Light*	England
39	36	1519	——	——	Wisby	Gothland
39	37	1520	——	——	Wismar	Mecklenburg
39	38	1521	——	——	Woahu, *Island*	Sandwich Isles

COMPOUND TABLE.

Pair Signals. No.	No.	Numerals.	Nations, Countries, Seas, Seaports, Capes, &c.	
40	1	1522	Wolf Rock	Moluccas
40	2	1523	Wollin, *Island*	Pomerania
40	3	1524	Wollsacken, Great	Pomerania
40	4	1525	Wollsacken, Little	Pomerania
40	5	1526	Wrangeroog, *Island, Light*	Oldenborg
40	6	1527	Wrath, *Cape, Light*	Scotland
40	7	1528	Xicoko	Japan
40	8	1529	Ximo	Japan
40	9	1530	Yarmouth	England
40	10	1531	Yeu, d', *Island, Light*	France, W
40	11	1532	York, *Cape*	New Holland
40	12	1533	Ystad	Sweden
40	13	1534	Zante, *Island*	Ionian Islands
40	14	1535	Zanzibar, *Island*	Africa, E
40	15	1536	Zelania, *Cape*	Nowaja Semlia
40	16	1537	Zeyla	Africa, E
40	17	1538	Zuider Zee	Holland
40	18	1539	Æroe, *Island*	Denmark
40	19	1540	Oeland, *Light*	Baltic
40	20	1541	Sound, The	Denmark
40	21	1542	Orkney, *Islands*	Scotland, N
40	22	1543	Oerskœr, *Light*	Gulf of Bothnia
40	23	1544	Oesel, *Island*	Baltic
40	24	1545	Oestergarnsholm, *Light*	Baltic
40	25	1546	Oester Riisoer	Norway
40	26	1547	Baltic, The	

II.

SELECTED SIGNALS,

BEING CERTAIN SIGNALS SET APART FOR SOME OF THE MOST IMPORTANT SENTENCES, REQUIRED TO BE COMMUNICATED WITH AS LITTLE LOSS OF TIME AS POSSIBLE.

THE SELECTED SIGNALS

are Single Signals, meaning thereby that each sentence is expressed by One Signal only, and not as in the Greater Code by two successive signals (or Pair Signals). The sentences have been selected, with a view to the importance of having the intelligence expressed by them communicated as quickly as possible.

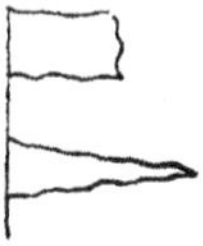

What is your latitude up to the last hour?
Ans. From Compound Table.

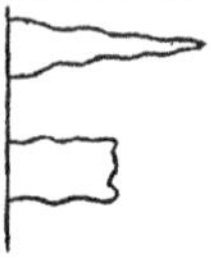

What is your longitude up to the last hour?
Ans. From Compound Table.

What are your *bearings* of the land, and *distance in leagues* from it; and what is the *name* of that land or place?
Ans. From Compound Table, with three pair signals in the same order as the questions follow.

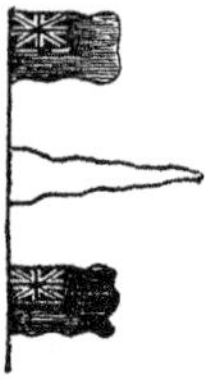

I see no possibility of keeping my vessel afloat, or of extinguishing the fire; save my crew!

If you continue your present course, you will be on shore, or run into other danger; haul to starboard (port your helm) till my next pair signal, from Compound Table, indicates the course you must steer.

If you continue your present course, you will be on shore, or run into other danger; haul to larboard (starboard your helm) till my next pair signal, from Compound Table, indicates the course you must steer.

Note.—If close hauled when the signal is made for luffing, it will of course be necessary to go about. And, to avoid the dangers alluded to, the safest plan will generally be to come round to the opposite course, until the next signal has pointed out the course to be steered.

War has been declared between the nations indicated next, by pair signals from the Compound Table.

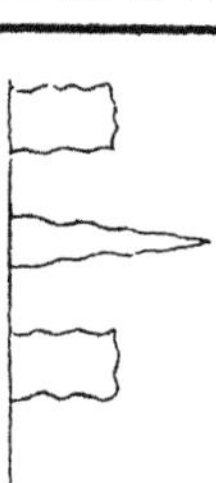

Peace has been concluded between the nations indicated next, by pair signals from the Compound Table.

III.

THE

LESSER CODE OF SENTENCES,

EXPRESSED BY

PAIR SIGNALS,

FOR THE USE OF VESSELS NOT BEING PROVIDED WITH THE NUMBER OF FLAGS REQUIRED FOR USING THE GREAT CODE.

INTRODUCTORY REMARKS.

Though it has been presumed that most vessels are provided with the flags required for the Great Code of Signals, yet, from accident or otherwise, cases may certainly occur, when there is only one national flag on board, or even no flag at all. We may suppose the party wishing to signalize to be so situated, as to have merely a bit of dark stuff for a coloured flag, and a couple of table-cloths, sheets, or even merely a couple of shirts, of whatever colour, instead of white flags. That a vessel even with such scanty means might be enabled to make known her wants, or communicate some important intelligence, this Lesser Code has been formed for universal use, upon the same principle of pair signals as the Great Code. The signals in this Code are arranged in such a manner as never to interfere with the signals of the Great Code.

The Preparatory Signal for this Code, indicating that "the signals following refer to the Lesser Code," or meaning, "I am obliged to make use of the Lesser Code for want of flags," is as represented underneath:

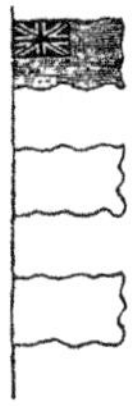

Note.—This signal also occupies a place in the Signal Table, having as such no other signification than its number and as a member of the pair-signal. It is only when hoisted the first time when entering upon a conversation, or an exchange of signals, that it conveys the above-mentioned meaning as Preparatory Signal.

SIGNAL TABLE.

Either the One

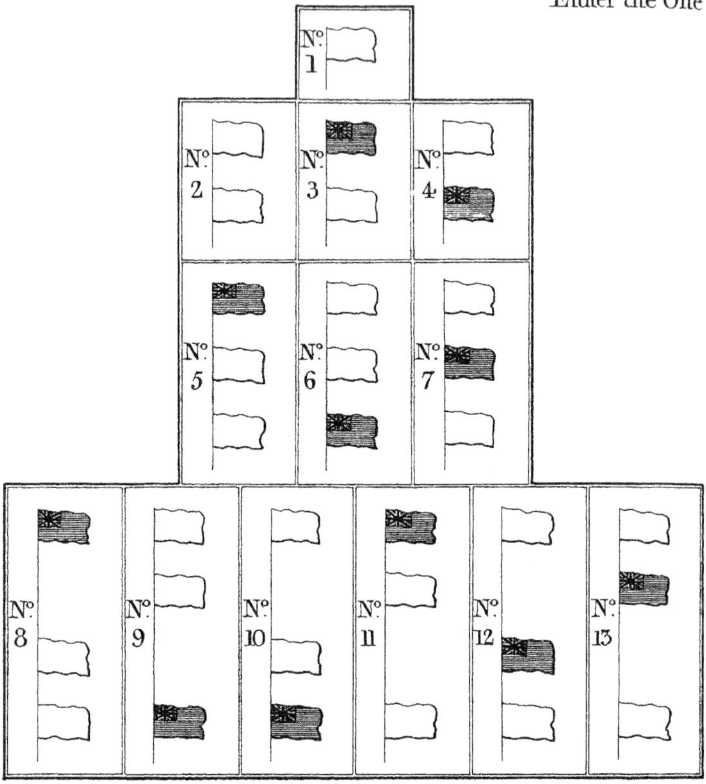

From the above Representation will be seen the peculiar mode which has been adopted in order to augment the number of Signals in the Table, the last six numbers being formed by a repeated combination of the three flags, with this difference, that a considerably greater distance is left open between either the uppermost or the lowermost flag and the next, than between the other two.— This distance ought not to be less than three times the usual one, in order to be conspicuous.

or the Other

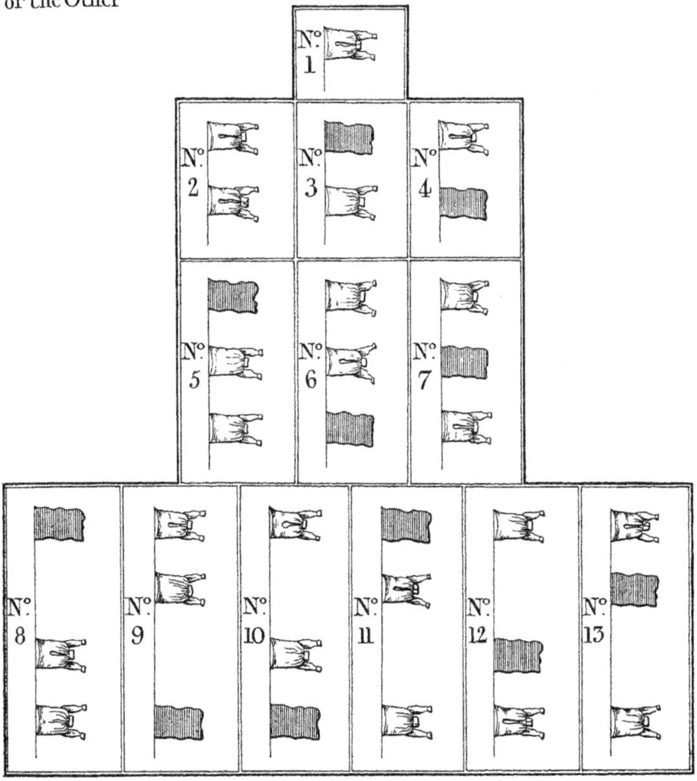

The above Table represents the Signals with shirts of any colour, instead of white flags, which are used in the opposite table.

Standidge & Lemon liths 77 Cornhill. London

The combination of the numbers in the signal table into pairs is the same as in the Great Code, and a sentence is expressed by two successive signals, or, as it is called for shortness' sake, by a pair-signal.

To this Code is likewise added a Compound Table, containing only three columns, viz.:

1.—For Numbers.
2.—For Points of the Compass.
3.—For Names of countries, seaport towns, headlands, &c.

And the pair signals for this table are the same as have been used to express the sentences. As has been explained in the introduction to the Compound Table for the Great Code, no mistake can arise as to whether a pair signal made refers to the Code of Sentences or to the Compound Table, from the same reason that the Compound Table is *never to be consulted but where a preceding sentence expressly indicates it.* Further directions for the use of this Compound Table will be given in another place.

The letters of the Alphabet are placed at the end of the sentences, and proper pair signals are assigned to them; and as these numbers are not used again in the Compound Table, they may be signalized without any previous signal for spelling, as no mistake can possibly arise about their meaning.

A signal is answered, or signified to be understood, by the hoisting of a white flag, which is hauled down when the signal goes down.

"A man is overboard," is signalized in the same manner as in the Great Code, by hoisting and hauling down the colours or national flag, and continuing to do the same till it is observed.

It ought to be remarked, that in the first ten pair-signals, the second signal in the pair is only the first hoisted with the flags in an inverted order; so that after hauling down the first signal in the pair, the second may be immediately hoisted without a change of flags.

CONTENTS

OF THE

CHAPTERS OF THE LESSER CODE.

SENTENCES

OF THE

LESSER CODE OF SIGNALS.

CHAPTER I.

Numerical order of Sentences.	Pair Signals. No.	Pair Signals. No.	*Relating to Intelligence requiring immediate Communication, or concerning matters of most frequent occurrence.* SENTENCES.
1	3	4	You are running into danger: bear up.
2	5	6	You are running into danger, haul your wind, or tack about.
3	8	9	What is your latitude?
			Ans. From Compound Table.
4	10	11	What is your longitude?
			Ans. From Compound Table.
5	12	13	Where do you belong to, and what is the name of your vessel?
			Ans. From Compound Table, name of place, and the vessel's name by spelling.
6	4	3	Where do you come from, and what is captain's name?
			Ans. Name of place, from Compound Table, and captain's name by spelling.
			Note.—It is to be presumed that a captain will communicate any news of importance, without waiting to be asked for it.
7	6	5	War has been declared between the *nations* indicated next (per Compound Table).
8	9	8	Peace has been concluded between the *nations* indicated next (per Compound Table).
9	11	10	Try to save vessel or people in distress, as per *bearings* (from Compound Table).
10	13	12	Do all you can to save my vessel or crew.

CHAPTER II.

Numerical order of Sentences.	Pair Signals. No.	Pair Signals. No.	*Relating to Navigation and Manœuvre of Vessels.* SENTENCES.
11	1	2	Steer the *course* as per signal next (per Compound Table).
12	1	3	Haul a point to starboard.
13	1	4	Haul a point to larboard.
			Note.—If more than one point is to be indicated, the last signal of the pair is hauled down, and rehoisted, as often as the points to be indicated amount to in number.
14	1	5	I shall tack or wear. If at anchor, this signal signifies "I shall weigh."
15	1	6	Will you carry a light to-night?
16	1	7	I shall carry a light to-night.
17	1	8	Will you bring to for the night?
			Ans. "Yes," or "No."
18	1	9	I shall bring to, during the night, from *hour* indicated next (from Compound Table, number of hours) till daylight.
19	1	10	Haul your wind upon the starboard tack.
20	1	11	Haul your wind upon the larboard tack.
21	1	12	Bear up, and steer large.
22	1	13	Will you lead during the night?
			Ans. "Yes," or "No."
23	2	1	Will you keep close to me to-night?
24	2	3	Shorten sail, I cannot keep up, or I wish to speak with you.
25	2	4	Do you intend to sail to-day?
			Ans. "Yes," or "No."

CHAPTER III.

Numerical order of Sentences.	Pair Signals. No.	No.	*Relating to Land, Shoregoing, Harbours, Entrances, &c.* SENTENCES.
26	2	5	Have you had sight of the land? *Ans.* "Yes," or "No."
27	2	6	I have seen the land bearing as indicated next (per Compound Table, point of compass).
28	2	7	Land is in sight *bearing* as per signal next (per Compound Table).
29	2	8	What are the bearings and distance in leagues from the nearest land by your reckoning? *Ans.* From Compound Table, by point of compass signal, and another pair signal for number of leagues.
30	2	9	Are you going into port? *Ans.* "Yes," or "No," or by the sentence following.
31	2	10	I am going into port, will you lead?
32	2	11	If you will follow me I will lead.
33	2	12	I am unacquainted with the harbour.
34	2	13	Do you take a pilot?
35	3	1	Have you got a pilot on board?
36	3	2	Is there much ice in the bay, or the passage, or entrance. *Ans.* "Yes," or "No," or by the sentence following.
37	3	5	The ice is on the side indicated next (per Compound Table, point of compass).
38	3	6	Are you acquainted with the entrance, or coast? *Ans.* "Yes" or "No," or by the following sentence.
39	3	7	I am not acquainted with the entrance, or coast, will you lead? *Ans.* "Yes," or "No," or by No. 33.
40	3	8	How many feet water is there in the channel you came by? *Ans.* From Compound Table in numbers.
41	3	9	I shall try to come to an anchor in shelter of the shore. Will you follow? *Ans.* "Yes," or "No."
42	3	10	I shall try to get nearer in shore. Will you follow? *Ans.* "Yes," or "No."

Numerical order of Sentences.	Pair Signals.		CHAPTER III. *Relating to Land, Shoregoing, Harbours, Entrances, &c.*
	No.	No.	SENTENCES.
43	3	11	I shall stand out to sea. Will you follow? *Ans.* "Yes," or "No."
44	3	12	Will you go on shore with me? I can offer you a passage in my boat? *Ans.* By No. 121 or 122.
45	3	13	I shall go on shore with you, if you will allow me a passage in your boat. *Ans.* By No. 121 or 122.

CHAPTER IV.

Numerical order of Sentences.	Pair Signals. No.	No.	*Relating to Average, Damage, Assistance.* SENTENCES.
46	4	1	My vessel has sprung a leak, will you keep near me ? *Ans.* "Yes," or according to circumstances.
47	4	2	I am aground, will you come to my assistance? *Ans.* As above.
48	4	5	I have sprung a mast.
49	4	6	I have sprung a topmast, can you spare me one? *Ans.* "Yes," or "No."
50	4	7	I am in distress, will you take me in tow, and bring me to the nearest port, or as far as you can ?
51	4	8	I have lost an anchor, can you spare me one ? *Ans.* "Yes," or "No."
52	4	9	I am short of water, can you spare me some ?
53	4	10	I am short of bread, can you spare me some ?
54	4	11	I am short of rum or spirits, can you spare me some ?
55	4	12	I am short of wood or coals, can you spare me some?
56	4	13	I am short of candles, can you spare me some ?
57	5	1	I am short of lamp oil, can you spare me some?
58	5	2	I am short of provisions, can you spare me some?
59	5	3	I am short of the article, or articles which shall be indicated next by spelling, "Can you spare me some?"
60	5	4	I have but little left of the article you are in want of, and nothing to spare.
61	5	7	Send your boat for the article wanted, or what I can spare.
62	5	8	Can you lend me your boat? *Ans.* "Yes," or by the sentence following.
63	5	9	My boat is already employed, or in a state not fit to be used.

CHAPTER V.

Numerical order of Sentences.	Pair Signals. No.	No.	*Relating to Commerce.* SENTENCES.
64	5	10	How is the market at the place you are coming from?
65	5	11	Market is good in general, or for article indicated next. *Note.*—If "in general" is to be understood, the last signal of the pair, when answered, is hauled nearly down, rehoisted, and struck.
66	5	12	Market is bad in general, or for article indicated next. *Note.*—Same note as above.
67	5	13	Does article indicated next find a good market at the place you are coming from? *Ans.* By "Yes," or "No," or by one of the two preceding sentences. *Note.*—If the article is not to be found here among the sentences, it must be expressed by spelling.
68	6	1	Corn.
69	6	2	Iron.
70	6	3	Timber.
71	6	4	Wine.
72	6	7	Fruits.
73	6	8	Coffee, sugar, or rum. *Note.*—The first of these articles is expressed by simply making the pair signal, and hauling it down in the ordinary way when answered; but to indicate sugar, the last signal of the pair when answered, is hauled nearly down, rehoisted, and struck; and when this is done twice over, it signifies rum.
74	6	9	West Indian produce.
75	6	10	Dry goods.
76	6	11	Provisions.
77	6	12	Stockfish.
78	6	13	Naval stores.
79	7	1	Are freights to be had at the place you are coming from? *Ans.* "Yes," or "No."
80	7	2	Good freights are to be had at the place I am coming from.

CHAPTER VI.

Numerical order of Sentences.	Pair Signals. No.	No.	Relating to Political and Warlike Events. SENTENCES.
81	7	3	At the place I am coming from, or *place* indicated next (per Compound Table) embargo has been laid on vessels of nation or nations indicated hereafter (per Compound Table). *Note.*—If "place coming from," is to be understood, the last signal of the sentence, when answered, is hauled nearly down, rehoisted and struck. After having thus indicated the place or country where an embargo has been laid on, the nation or nations on whose vessels the embargo has been laid are next signalized from the Compound Table. If there be one nation only, the last signal of the pair, when answered, is hauled nearly down, rehoisted, and struck, to indicate that no more signals for nations are to come; and if there be two or more nations to be signalized, this hauling down and rehoisting the terminating signal is only observed with the last of them.
82	7	4	At the place I am coming from, or *place* indicated next (per Compound Table), embargo has been taken off the vessels of nation or nations indicated hereafter (per Compound Table). *Note.*—The preceding note applies equally to this sentence.
83	7	5	The harbour or harbours of the *country* indicated next (per Compound Table), is or are blockaded by the *nation* or *nations* indicated by the second or following pair signals:—(per Compound Table.) *Note.*—The harbour or country blockaded is first signalized, and after that follows the signal for nation or nations blockading, for which see note to No. 81.
84	7	6	Blockade has been taken off *place* indicated next (per Compound Table).
85	7	8	Have you seen enemy's vessels during your passage? *Ans.* "No," or by sentence following.
86	7	9	I have seen enemy's ships *steering* as indicated next (per Compound Table).
87	7	10	I have met with or seen privateers of *nation* indicated next (per Compound Table).
88	7	11	Did you on your passage fall in with or see any man-of-war? *Ans.* By "No," or by sentence following.
89	7	12	I have met with or seen men-of-war of the *nation* indicated next (per Compound Table).
90	7	13	I have seen men-of-war, but cannot tell what nation they were of.

CHAPTER VII.

Numerical order of Sentences.	Pair Signals. No.	No.	*Relating to Inquiries about Captains, Passengers, State of Health, Home, Voyage, and of other Vessels.* SENTENCES.
91	8	1	Are all well on board?
92	8	2	All are well.
93	8	3	Captain or a passenger is sick (passenger's name to be spelled).
94	8	4	Captain, or a passenger, died on the passage (if passenger, or another person than captain, name will be spelled).
95	8	5	Have you passengers on board? *Ans.* By "Yes," or "No."
96	8	6	How are passengers?
97	8	7	I have a passenger, or passengers wishing to go with you; can you accommodate him, her, or them? *Number* to be indicated next (per Compound Table). *Ans* By "Yes," or "No," or by the following sentence: *Note.*—If only one passenger, the last signal of the sentence, when answered, is hauled nearly down, rehoisted, and struck.
98	8	10	I can accommodate passengers, *number* indicated next (per Compound Table).
99	8	11	I have sick on board, and am in want of medicine, can you spare me some?
100	8	12	I have, or vessel indicated as per *bearings* (from Compound Table) has contagious disease on board. *Note.*—If "I have," is to be understood, the last signal of the sentence, when answered, is hauled nearly down, rehoisted, and struck.
101	8	13	Contagious disease or the plague, is raging at place I am coming from, or *place* indicated next (per Compound Table). *Note.*—If "place coming from" is to be understood, the last signal of the sentence, when answered, is hauled nearly down, rehoisted, and struck.
102	9	1	Quarantine law has been put in force at the place I am coming from, or *place* indicated next (per Compound Table). *Note.*—If "place coming from" is to be understood, the last signal of the sentence, when answered, is hauled nearly down, rehoisted, and struck.
103	9	2	Are my family well at home? *Ans.* By No. 92, or by sentence following.

Numerical order of Sentences.	Pair Signals. No.	No.	CHAPTER VII. *Relating to Inquiries about Captains, Passengers, State of Health, Home, Voyage, and of other Vessels.* SENTENCES.
104	9	3	I do not know.
105	9	4	Report my ship when you arrive.
106	9	5	Have you any late newspaper?
107	9	6	Heave to! I will send letters on board.
108	9	7	How many days have you been out? *Ans.* By numbers, Compound Table.
109	9	10	Were there many vessels of my nation at the place you are coming from? *Ans.* "Yes," or "No."
110	9	11	Had vessel belonging to *place* indicated next (per Compound Table), and whose name shall be spelled, arrived when you left?
111	9	12	Where does vessel hailed belong to, and what is her name? *Ans.* Place or country, from Compound Table, and name of vessel by spelling.
112	9	13	Where does vessel hailed come from, and what is the name of her captain? *Ans.* Name of place from Compound Table, and captain's name by spelling.

CHAPTER VIII.

Numerical order of Sentences.	Pair Signals.		*Relating to Pirates, and Overhauling by Privateers.*
	No.	No.	SENTENCES.
113	10	1	Have you heard of or seen pirates or suspicious vessels? *Ans.* "Yes," or "No."
114	10	2	Vessel, or vessels, *bearing* as indicated next (per Compound Table), is, or are, suspicious.
115	10	3	My vessel, or another vessel, has been plundered in *latitude* and *longitude* indicated next (per Compound Table). *Note.*—If "my vessel" is to be understood, the last signal of the sentence when answered, is hauled nearly down, rehoisted, and struck.
116	10	4	My vessel, or another vessel, has been overhauled by a privateer of *nation* indicated next (per Compound Table), in *latitude* and *longitude* indicated hereafter (both from Compound Table). *Note.*—If "my vessel" is to be understood, the last signal of the sentence, when answered, is hauled nearly down, rehoisted, and struck.
117	10	5	Let us keep company till we have passed the place suspected, and keep a good look out for suspicious vessels.

CHAPTER IX.

Numerical order of Sentences.	Pair Signals.		*Miscellaneous.*
	No.	No.	SENTENCES.
118	10	6	What news from the places you left?
119	10	7	I have letters or things for you, send for them.
120	10	8	Will you, or person whose name shall be spelled, come on board?
121	10	9	With pleasure.
122	10	12	Beg to be excused.
123	10	13	Will you breakfast, dine, or sup with me?
			Note.—If the signal be made before 8 o'clock A.M. it means "breakfast;" and after 4 o'clock P. M., the invitation is meant for "supper."
124	11	1	I cannot distinguish the signal flying.
125	11	2	YES.
126	11	3	NO.
127	11	4	Signal made (that is to say, the meaning of it), not understood, although the flags were distinguished.

THE ALPHABET.

Pair Signals.			
11	5	(')	Sign of Abbreviation
11	6	A	
11	7	B	
11	8	C	
11	9	D	
11	12	E	
11	13	F	
12	1	G	
12	2	H	
12	3	I	
12	4	J	
12	5	K	
12	6	L	
12	7	M	
12	8	N	
12	9	O	
12	10	P	
12	11	Q	
13	1	R	
13	2	S	
13	3	T	
13	4	U	
13	5	V	
13	6	W	
13	7	X	
13	8	Y	
13	9	Z	
13	10	Æ	
13	11	Œ	

DIRECTIONS

FOR THE USE OF THE COMPOUND TABLE

OF THE LESSER CODE.

The *Column for Numerals* serves to express the number of every thing. The corresponding pair signals represent all the numbers from one to a hundred, all the hundreds to one thousand, and the thousands from one to seventeen thousand, by which signals any given number not exceeding 17,999, may be expressed in the following manner. For example: 984 is to be expressed by signals; then hoist the pair signal No. 9 and 7, for nine hundred, and after that the pair signal No. 7 and 6, for eighty-four. If the number to be expressed were 17,999; hoist pair signal No. 11 and 2, for seventeen thousand, then the pair signal No. 9 and 7, for nine hundred, and lastly, the pair signal No. 8 and 11, for ninety-nine.

The terminating signal of the last pair signal (or if only one pair is required, then, the terminating signal of that pair) is, when answered, to be hauled nearly down, rehoisted, and struck, in order to indicate that the number is full.

The *Latitude* and *Longitude* are expressed by this same column; it being taken for granted, that a preceding sentence signal has either implied, that latitude and longitude were to follow, or has requested the one or the other to be communicated. The number of degrees is first signalized, and next the number of minutes. If in the preceding sentence mention has been made of both latitude and longitude; then, the first two pair signals refer to the number of degrees and minutes for *latitude*, and the following pair signals express the degrees and minutes of *longitude*, and it must be remembered that as the degrees of the longitude may require more than one pair of signals, the minutes must not be signalized, until the above-mentioned operation with the rehoisting, and hauling down of the last signal, has been observed. The cipher (0) will be found in the same column, placed after all the numbers; it is signalized where 0° or 0′ occur in the latitude or longitude; as for example; the latitude to be signalized is 0° 26′ North; then hoist the pair signal No. 11 and No. 3, for 0°, and next the pair signal No. 2 and No. 5, for 26′; or, let the longitude to be signalized be 36° 0′; then hoist the pair signal No. 3 and No. 2, for 36°, and next the pair signal No. 11 and No. 3, for 0′. And in very low latitudes, in order to avoid all mistakes, we will suppose the possibility of so great an error in the reckoning as to make it doubtful whether North or South is meant), then the pair signal indicating the letter N may be hoisted, after the minutes have been signalized, when the latitude is north, and the letter S when the latitude is south.

The longitude is reckoned from the meridian of Greenwich, and constantly east from 0° 0′ to 359° 59′. But, for brevity's sake, the pair signal for the hundreds may in many cases be omitted; as for example: being in longitude 285° by observation, to communicate this, it will only be necessary to make the signals for 85°, as the other party may from his own reckoning supply the hundreds as a matter of course.

The column for *Points of the Compass*, contains the bearings to every half point, all round the compass. The signals only refer to this column, where the preceding sentence signals expressly have mentioned it.

Countries, *Seaport Towns*, and some of the principal *Headlands* will be found in the last column of the Table, which also is only to be looked to at the express mention of it in the preceding sentence signal.

For a fuller explanation of the nature and principles of this Compound Table, see Directions for Compound Table to the *Great Code of Signals*.

COMPOUND TABLE OF THE LESSER CODE.

Pair Signals.		Numerals.	Points of the Compass.	Countries, Seaports, Headlands, &c.
3	4	1	North	Africa, *Coast of*
5	6	2	N ½ E	Alexandria. EGYPT.
8	9	3	N by E	Algiers.
10	11	4	N by E ½ E	Altona.
12	13	5	NNE	Amsterdam. South Point of Texel.
4	3	6	NNE ½ E	Anegada.
6	5	7	NE by N	Antigua.
9	8	8	NE ½ N	Antwerp. West Caple.
11	10	9	NE	Archangel.
13	12	10	NE ½ E	Arendal.
				AUSTRIA. See 11.3
1	2	11	NE by E	Baltimore.
1	3	12	ENE ½ N	Barbadoes.
1	4	13	ENE	Bengal.
1	5	14	E by N ½ N	Bergen in Norway.
1	6	15	E by N	Boston in America.
1	7	16	E ½ N	Bordeaux. Cordouan Light.
1	8	17	E	BRAZIL.
1	9	18	E ½ S	Bremen.
1	10	19	E by S	Brest. Ushant.
1	11	20	E by S ½ S	Bristol. Worm's Head.
1	12	21	ESE	Cadiz. *Cape* St. Vincent.
1	13	22	ESE ½ S	Canary Islands.
2	1	23	SE by E	Canton. CHINA.
2	3	24	SE ½ E	Cape of Good Hope.
2	4	25	SE	Christiania. Færder.
2	5	26	SE ½ S	Christiansand.
2	6	27	SE by S	Constantinople. TURKEY.
				Copenhagen. See 5.9
2	7	28	SSE ½ E	Cork. IRELAND.
2	8	29	SSE	Cronstad. St. Petersburg.
2	9	30	S by E ½ E	Corsica.
2	10	31	S by E	DENMARK.
2	11	32	S ½ E	Dantzic.
2	12	33	S	Dover. South Foreland.
2	13	34	S ½ W	Dram.
				Drontheim. See 10.
3	1	35	S by W	Dunkirk.
				EAST INDIES. See 7.8
3	2	36	S by W ½ W	Edinburgh. SCOTLAND.
3	5	37	SSW	Eyder.
3	6	38	SSW ½ W	Elbe. Heligoland.
				Elsinore. See 4.13
3	7	39	SW by S	ENGLAND.

COMPOUND TABLE OF THE LESSER CODE.

Pair Signals.		Numerals.	Points of the Compass.	Countries, Seaports, Headlands, &c.
3	8	40	SW ½ S	FEROE ISLANDS.
3	9	41	SW	FALMOUTH.
3	10	42	SW ½ W	FERROL. *Cape* ORTEGAL.
3	11	43	SW by W	FLENSBURG.
3	12	44	WSW ½ S	FRANCE.
3	13	45	WSW	GIBRALTAR.
4	1	46	W by S ½ S	GLASGOW.
4	2	47	W by S	GOTHENBORG. WINGOE SAND.
4	5	48	W ½ S	GREENLAND.
4	6	49	W	GUADALOUPE.
4	7	50	W ½ N	GUERNSEY.
4	8	51	W by N	HALIFAX.
4	9	52	W by N ½ N	HAMBURG.
4	10	53	WNW	HARWICH. ORFORDNESS.
4	11	54	WNW ½ N	HAVANNAH.
4	12	55	NW by W	HAVRE DE GRACE.
4	13	56	NW ½ W	ELSINORE.
5	1	57	NW	HOLLAND.
5	2	58	NW ½ N	HULL. SPURN POINT.
5	3	59	NW by N	JAMAICA.
5	4	60	N N W ½ W	ICELAND.
5	7	61	NNW	ISLE DE FRANCE.
5	8	62	N by W½ W	ITALY.
5	9	63	N by W	COPENHAGEN.
5	10	64	N½ W	LISBON.
5	11	65	. . . .	LIVERPOOL.
5	12	66	. . .	LEGHORN.
5	13	67	. . . .	LONDON.
6	1	68	. . .	LUBEC. TRAVEMUNDE.
6	2	69	. . . .	MADEIRA.
6	3	70	. . . .	MALAGA.
6	4	71	. . . .	MALTA.
6	7	72	. . . .	MANDAL. THE NAZE (LINDERSNÆS).
6	8	73	. . . .	MARGATE. NORTH FORELAND.
6	9	74	. . . .	MARSTRAND, *Light*
6	10	75	. . . .	MARTINIQUE.
6	11	76	. . . .	MESSINA.
6	12	77	. . . .	NAPLES.
6	13	78	. . . .	NYBORG.
7	1	79	. . . .	NEWCASTLE. TYNEMOUTH CASTLE.
7	2	80		NEW YORK.
7	3	81		NORTH AMERICA. (U. STATES.)
7	4	82		NORWAY.
7	5	83		OPORTO.
7	6	84		OKNEY ISLANDS. SOUTH RUNDLE'S [HEAD.

COMPOUND TABLE OF THE LESSER CODE.

Pair Signals.		Numerals.	Countries, Seaports, Headlands, &c.
7	8	85	EAST INDIES.
7	9	86	PALERMO.
7	10	87	PHILADELPHIA.
7	11	88	PLYMOUTH. RAM'S HEAD.
7	12	89	PORTO-RICO.
7	13	90	PORTSMOUTH.
8	1	91	PORTUGAL.
8	2	92	PRUSSIA. MEMEL.
8	3	93	REVAL.
8	4	94	RIGA.
8	5	95	RIO DE LA PLATA.
8	6	96	RIO DE JANEIRO.
8	7	97	ROSTOCK.
8	10	98	ROTTERDAM. GORÉE.
8	11	99	RUSSIA.
8	12	100	SCAW, *Light.*
8	13	200	ST. ANDER.
9	1	300	ST. CROIX, WEST INDIAN ISLAND.
9	2	400	ST. DOMINGO, OR HAYTI.
9	3	500	ST. JOHN'S, NEWFOUNDLAND.
9	4	600	ST. THOMAS, WEST INDIAN ISLAND.
9	5	700	ST. UBES, OR SETUVAL.
9	6	800	SARDINIA.
9	7	900	SCARBOROUGH. FLAMBOROUGH-HEAD.
9	10	1000	SHETLAND, SOUTHERN END.
9	11	2000	SICILY.
9	12	3000	SPAIN.
9	13	4000	STETTIN. SVINEMUNDE.
10	1	5000	STOCKHOLM.
10	2	6000	STRALSUND.
10	3	7000	SWEDEN.
10	4	8000	SOUTH AMERICA.
10	5	9000	TOULON.
10	6	10,000	TRANQUEBAR.
10	7	11,000	TRIEST.
10	8	12,000	TRIPOLIS.
10	9	13,000	DRONTHEIM.
10	12	14,000	TUNIS.
10	13	15,000	VENICE.
11	1	16,000	WEST INDIES.
11	2	17,000	VIGO. CAPE FINISTERRE.
11	3	0	AUSTRIA.

IV.

SIGNALS TO BE USED BY PILOTS

TO

VESSELS IN DISTRESS,

WHEN IT BLOWS SO HARD THAT THEY CANNOT BE BOARDED.

The following 24 signals are to be used by pilots, or persons on shore, to give directions to a vessel in distress outside, to enable it to steer into a port or anchorage, when it blows too hard for a pilot to board it. These signals require two national or coloured flags, two white ditto, and a pendant; and though they appear, at first sight, to resemble some of the signals of the Lesser Code, on account of the same mode of multiplying the combination of the flags having been used in both, by leaving a distance between the upper or lowermost flag and the next, three times greater than that between the other two, yet they cannot be mistaken as belonging to that code, or any other signals previously set forth in this book; because neither the pendant, nor the two national or coloured flags, enter into the former one, and the distinguishing feature of these signals—the unequal space between the flags—is entirely wanting in all but the six last combinations of flags in the signal table of the Lesser Code. The vessel in distress seeing these signals hoisted on shore, or from a vessel in the harbour, may therefore immediately understand the signals as intended for her. They may be hoisted either from a pole or mast on shore, in a conspicuous place, or from the mast-head of a vessel in the harbour.

TABLE OF PILOT SIGNALS

No. 1 — Starboard your helm (turn to larboard) and continue turning until this Signal is hauled down, by which act, the word: "*steady*" is expressed.	No. 2 — Port your helm, (turn to starboard) and continue turning until this Signal is hauled down, by which act, the word "*steady*" is expressed.	No. 3 — Tack, if possible.
No. 7 — Let go all your anchors, and do all possible to ride it out. (cut away masts, if necessary)	No. 8 — Prepare to beach vessel, in the direction indicated hereafter.	No. 9 — Steer by Compass North. N.B. hoisted twice it signifies N. by E.
No. 13 — Steer by Compass East. N.B. hoisted twice it signifies E. by S.	No. 14 — Steer by Compass E. S. E. N B hoisted twice it signifies S.E. by E.	No. 15 — Steer by Compass S. E. N B hoisted twice it signifies S.E by S.
No. 19 — Steer by Compass S. W. N.B. hoisted twice it signifies S.W. by W.	No. 20 — Steer by Compass W. S. W. N B hoisted twice it signifies W. by S.	No. 21 — Steer by Compass West. N B hoisted twice it signifies W. by N.

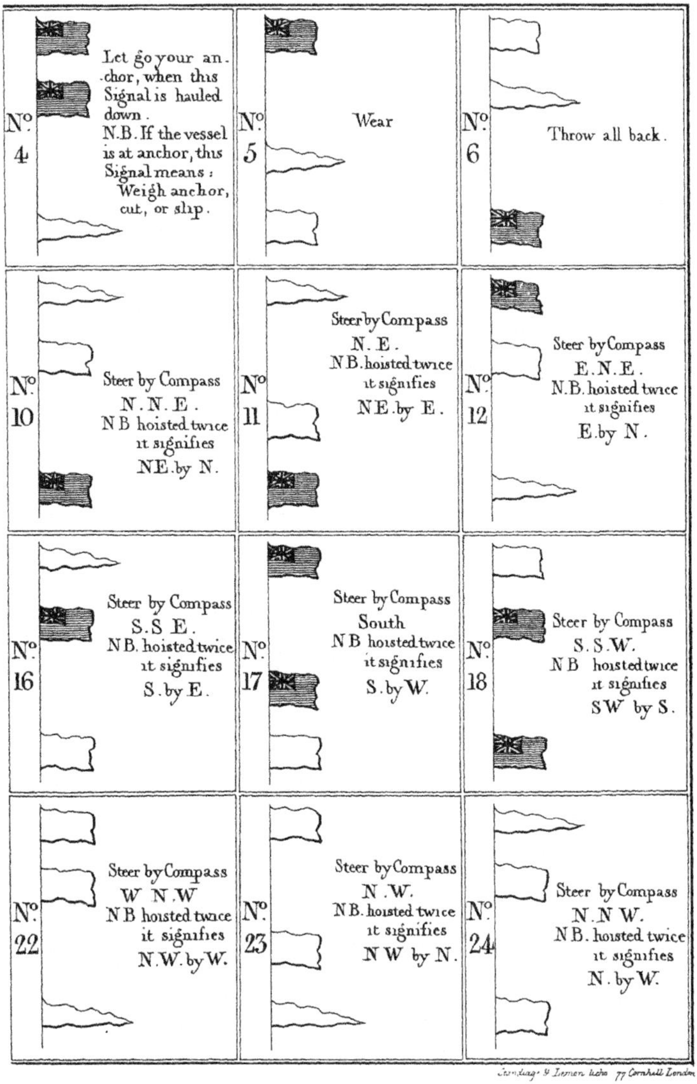

Standidge & Lemon litho 77 Cornhill London

GENERAL NOTE.

By a nearer inspection of the preceding Table, it will be remarked, that No. 2 is the signal No. 1 inverted: No. 4 is the same as No. 3, only hoisted with the flags in inverted order, and so forth; each of the even numbers having the same signal as the uneven number immediately preceding it, but in an inverted order. The sentences are arranged with a view to the advantage to be derived herefrom; as, for example: "Port your Helm," or "Turn to Starboard," is the inverted signal for "Starboard your Helm," or "Turn to Larboard." A vessel may thus be conned through an intricate passage by the one signal, without any change of flags being necessary; only hoisting it alternately the one way, or the other, as required. In the same manner, "Let go your Anchor," is the inverted signal for "Tack, if possible." Because if the vessel should miss stays, and want of room, or other circumstances, not allow of other measures, the advice most likely to be given would be to come to an anchor immediately. It is therefore necessary that the pilot, or person making signals, assure himself, before changing the flags to a new signal, that the signal to be made next, be not the same as the cne just hauled down, with only the flags in an inverted order.

It is recommended to be provided with a loose tack to the flag or pendant, which is to be at triple distance from the others; this tack ought to have the length of somewhat more than three times the width or hoist of one of the signal flags; and if provided with a small spring hook in one or both ends, the fastening of the flags at a proper distance will be greatly facilitated.

V.

SIGNALS

TO BE USED BY VESSELS HAVING LOST THEIR MASTS, OR FOR WRECKS.

A VESSEL having met with the misfortune of losing all her masts, no spar being left standing to hoist a signal from, may still communicate some particular want, or point out any assistance particularly desired, by making use of the following signals; and though the helplessness of her situation sufficiently bespeaks her want of immediate assistance, which no doubt will be cheerfully rendered her, by every one who is able to do so without waiting for a positive demand, yet there may be cases where some of the following signals will be found useful.

These signals are made by holding up poles or sticks with the colours, and two white flags, sheets, or table cloths fastened to them: a dark-coloured handkerchief and two white ones, or a handkerchief or other piece of whatever colour (white not excepted), and two shirts may be substituted: it matters not what colour the two last-mentioned be of, provided they may be distinguished for what they are. It is of course taken for granted that the decks are cleared, and that nothing hinders the signals from being clearly seen.

The flags are held up, one, two, or three at a time, by as many men placed along deck, at proper distances from each other, and they must be *read*, as shown in the following representations of the signals, from *left to right:* it is therefore necessary on board the vessel to consider what flag will appear on the left-hand to the persons for whom the signals are making; and if there happen to be people on both sides of the vessel, from whom assistance may be expected, the left flag is waved repeatedly, and again held steadily up, to indicate which flag is to be considered as the *first* or *left flag* by both parties.

TABLE OF SIGNALS FOR VESSELS HAVING

I Representation.		
N° 1	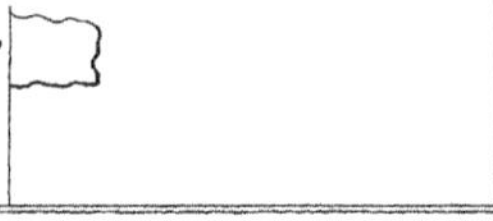	Signal made is understood.
N° 2	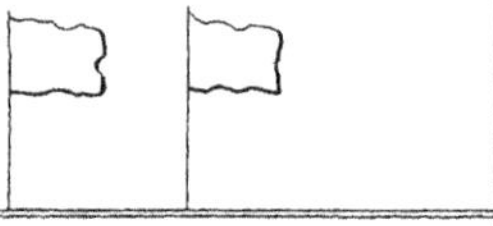	Signal made is not understood, repeat the same.
N° 3	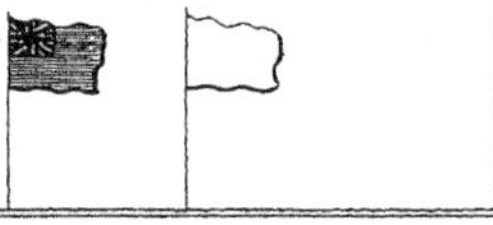	Take me in tow, or save my vessel and crew. *
N° 4		I am in the greatest distress; send provisions.
N° 5	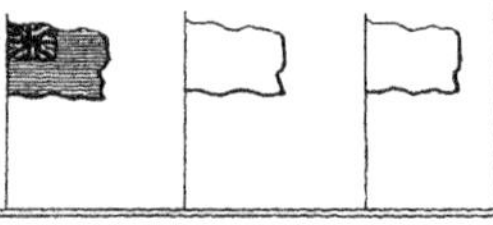	I am in want of anchor and cable; endeavour to bring me them.
N° 6	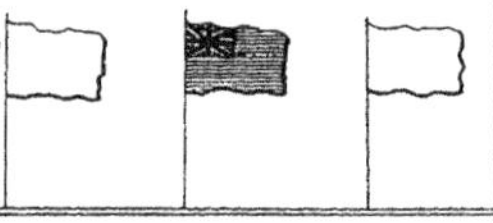	I am in great distress, send provisions, and an anchor and cable.
N° 7	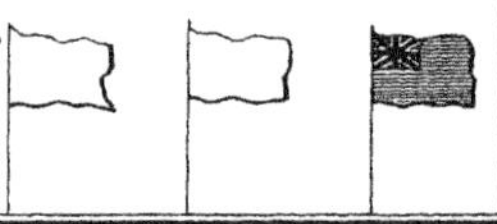	Shew by the waving of a flag, or other-wise point out, where I may try to beach the wreck, if possible.

* Should it so happen that there is only one person remaining on board, he must make fast the right hand Pole, or Stick, with its flag; and then, by himself holding the left hand one, he will be able to make the only signal which he naturally would wish to make; viz, to save him.

NO MASTS STANDING, OR FOR WRECKS.

II Representation.		
Nº 1		Signal made is understood
Nº 2	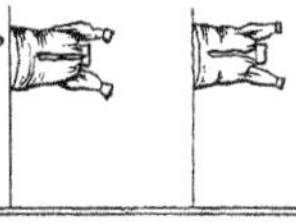	Signal made is not understood, repeat the same.
Nº 3	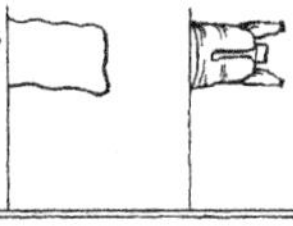	Take me in tow, or save my vessel and crew. *
Nº 4	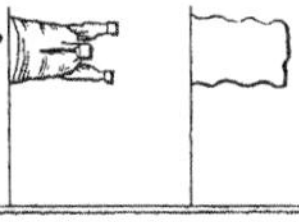	I am in the greatest distress ; send provisions.
Nº 5	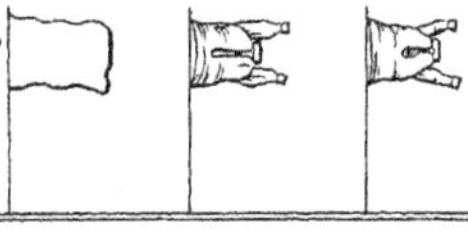	I am in want of anchor and cable ; endeavour to bring me them.
Nº 6	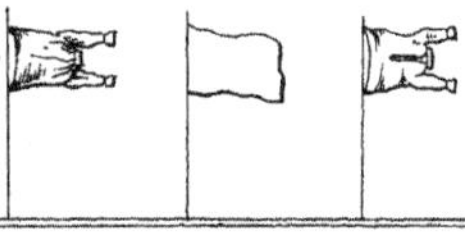	I am in great distress ; send provisions, and an anchor and cable.
Nº 7		Shew by the waving of a flag, or other-wise point out, where I may try to beach the wreck, if possible.

It has been thought necessary to give the above Representation for the sake of greater clearness, in cases when a Handkerchief of any colour, and two Shirts, also of any colour, are to be used instead of Flags.

VI.

NIGHT SIGNALS.

A SMALL yard, eight feet long, must be rigged for the night signals, in the following manner: At one foot from each end of the yard-arm are fastened the ends of a small rope, having twice the length of the yard, forming a crow-foot with a thimble in the bight. Two such ropes form the rigging of the signal yard, one to show upward, the other downward; and in the thimble in the bight of the former, is fastened the signal haliard; and in that of the latter the downhaul, as represented in the sketch underneath.

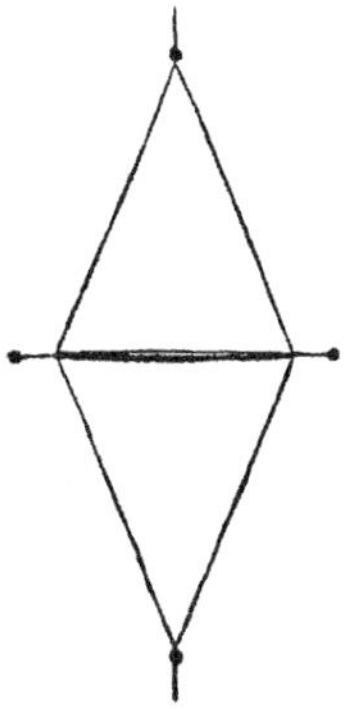

This yard and three lanterns form the whole apparatus of these night signals. When three lanterns are to be hoisted from the yard, they may either represent a triangle, to the one or the other side of the yard, the top or vertex, being either upward or downward; or they may be arranged in a horizontal line along the yard, when a small iron ring, or eye, at either end of the yard-arm, and a notch cut round the middle of the yard, serve to fasten the lanterns in their proper places. Two horizontal lanterns are fastened, one on each end of the yard.

Two or three lanterns, in a vertical line, are hoisted simply by the signal haliard, with tacks of from 5 to 6 feet long between.

With the three lanterns, only 7 signals can be made, as represented in the Signal Table, page 165; and as this very limited number would be quite insufficient for expressing the number of sentences deemed necessary, the system of pair signals has likewise been adopted for this Code of Night Signals, which consists of 36 sentences. Before each sentence are placed the numbers of the two successive signals by which it is expressed. By the combination in pairs, care has been taken that a pair signal should never begin with No. 1, or the single lantern, which, in the beginning of a signal, might be taken for another light accidentally shown, or *vice versâ.* In the same manner, will be found omitted, all the pairs composed of the same figure, as No. 2, No. 2; No. 3, No. 3, &c., to avoid mistakes; as it might happen, that by the rolling of the vessel, the lanterns of the first signal were hid for a moment behind the sails or rigging, and might at their re-appearance be taken for the second signal of the pair.

SIGNAL TABLE.

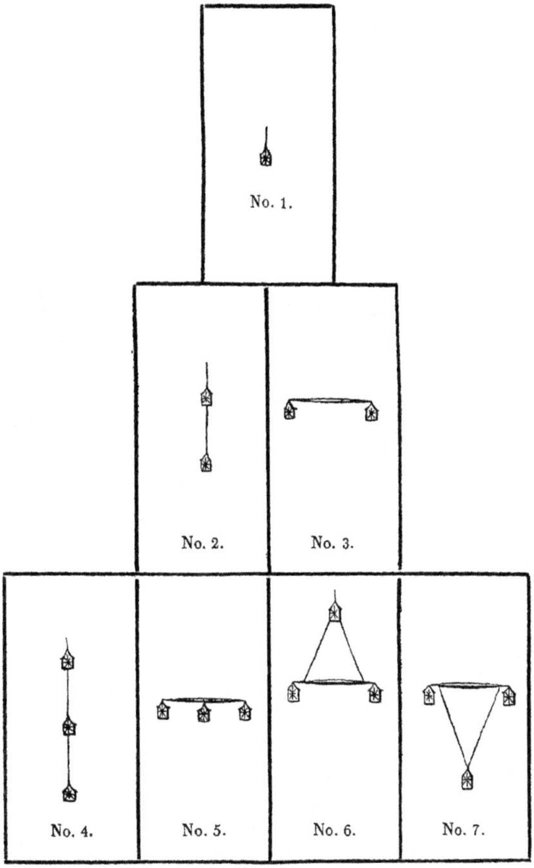

By totally excluding the use of guns for signals, this Code becomes fit to be used by smaller vessels, which commonly are not provided with a gun; and another advantage is, that the noise, occasioned by the firing of guns, is avoided; which, in the presence of an enemy, cannot fail to attract notice, and may be the cause of discovery and capture. When otherwise advisable and necessary to call the attention to a signal, a gun may be fired, without altering the sense of the signal or signals; and this may likewise be reckoned an advantage peculiar to this system.

Before going to signalize, or enter into conversation, it is necessary to make a preparatory signal, which cannot be mistaken for another, or other lights shown. This preparatory signal consists in hoisting *two vertical lanterns*, thus :

when at the same time, from another place, is hoisted *a single lantern*, hauling the same up and down, until observed and answered, from the other vessel in the same manner.

When the signalizing is going on, each single signal is answered, by hoisting a single lantern, which is hauled down, when the signal is taken in. And care must be taken to put out, or cover up, all other lights on board, as long as the signalizing is going on.

NIGHT SIGNALS.

Numerical order of Sentences.	Pair Signals. No.	Pair Signals. No.	*Night Signals.* SENTENCES.
1	2	1	A man is overboard; save him if you can.
2	2	3	Signal made not understood; repeat the same.
3	2	4	Stand not too far on the present tack.
4	2	5	Yes.
5	2	6	No.
6	2	7	I wish to speak with you; come within hail.
7	3	1	My vessel is leaky; keep near me.
8	3	2	I cannot keep my vessel from sinking; save my crew.
9	3	4	Fire is broken out in my vessel; keep with me, and come to my assistance.
10	3	5	I am ashore; come to my assistance.
11	3	6	I see land, or light on shore, as per bearing indicated next (point of compass, per Compound Table).
12	3	7	What distance do you reckon yourself from land?
			Ans. By a pair signal from the annexed Compound Table, for distance in cable's lengths.
13	4	1	I shall tack when this signal goes down.
14	4	2	Stand away from me, or you'll be on shore as I am.
15	4	3	I shall wear when this signal goes down.
16	4	5	I am steering the course as indicated next (points of compass, per Compound Table).
17	4	6	You'll do best to steer the course as indicated next (points of compass, per Compound Table).
18	4	7	Show the signal for going about, when you are going to tack, or wear.
19	5	1	I am about; and steering the course indicated next (point of compass, per Compound Table).
20	5	2	Upon what tack are you?
			Ans. By one of the two following sentences, or by sentence No. 16.
21	5	3	I am upon the starboard tack.
22	5	4	I am upon the larboard tack.
23	5	6	I am going to anchor.
24	5	7	I am come to an anchor.
25	6	1	I shall weigh.
26	6	2	I shall cut or slip, and stand out to sea.
27	6	3	I shall cut or slip, and stand for the nearest port.
28	6	4	I am under weigh.

Numerical order of Sentences.	Pair Signals. No.	No.	*Night Signals.* SENTENCES.
29	6	5	Carry a light to-night, and I will follow you.
30	6	7	I shall take the lead, and show a light, if necessary.
31	7	1	Shorten sail, that I may keep up.
32	7	2	Make sail, if you mean to follow.
33	7	3	Do you intend to go in to-night?
34	7	4	I intend to run in to-night.
35	7	5	I think it the best to separate, each doing his best to get away from here.
36	7	6	Let us keep together to-night, and keep a good look-out.

COMPOUND TABLE

FOR THE

NIGHT SIGNALS.

A Compound Table has been arranged to these Night Signals, similar to those to the preceding Codes: this Table contains only two columns, one for distance in cable's lengths, and another for points of the compass. (See the other side.)

COMPOUND TABLE FOR THE NIGHT SIGNALS.

Pair Signals.		Distance.	Points of the Compass.
2	1	1 Cable's length.	NORTH
2	3	1½ ditto	N by E
2	4	2 ditto	N N E
2	5	3 ditto	N E by N
2	6	4 ditto	N E
2	7	5 ditto	N E by E
3	1	6 ditto	E N E
3	2	7 ditto	E by N
3	4	8 ditto	EAST
3	5	9 ditto	E by S
3	6	10 ditto	E S E
3	7	11 ditto	S E by E
4	1	12 ditto	S E
4	2	13 ditto	S E by S
4	3	14 ditto	S S E
4	5	15 ditto	S by E
4	6	16 ditto	SOUTH
4	7	17 ditto	S by W
5	1	18 ditto	S S W
5	2	19 ditto	S W by S
5	3	20 ditto	S W
5	4	21 ditto	S W by W
5	6	22 ditto	W S W
5	7	23 ditto	W by S
6	1	24 ditto	WEST
6	2	25 ditto	W by N
6	3	26 ditto	W N W
6	4	27 ditto	N W by W
6	5	28 ditto	N W
6	7	29 ditto	N W by N
7	1	30 ditto	N N W
7	2	31 a 40 ditto	N by W
7	3	41 a 50 ditto	
7	4	51 a 60 ditto	
7	5	61 a 70 ditto	
7	6	71 and upwards ditto	

N.B.—30 Cable's lengths=1 League.

FINIS.

C. WHITING, BEAUFORT HOUSE, STRAND.

ALPHABETICAL INDEX

TO

SENTENCES IN THE GREAT CODE.

ERRATA.

Page 85,	Pair Signals,	2.11,	*for* ALCONA *read* ALTONA.
———	———	2.17,	*for* AMRAM *read* AMRUM.
—— 87,	———	4.37,	*for* *Flag Light* read *Floating Light*

Printed in Dunstable, United Kingdom

91219505R00117